Amaranth
from the past—for the future

Amaranth
from the past for the future

John N. Cole

Rodale Press, Emmaus, Pennsylvania

Book design by Barbara Field

Printed in the United States of America on recycled paper, containing a high percentage of de-inked fiber.

Library of Congress Cataloging in Publication Data

Cole, John N 1923-
Amaranth, from the past for the future.

Bibliography: p.
Includes index.
1. Amaranthus. 2. Food crops. 3. Plant lore.
I. Title.
SB191.A42C64 338.1'7'3 79-16953
ISBN 0-87857-240-6

2 4 6 8 10 9 7 5 3 1

Time present and time past
Are both perhaps in time future
And time future contained in times past.

T. S. Eliot—*Four Quartets*

Ending hunger on earth is an idea
whose time has come.

Andrew Young—U.S. Ambassador
to the United Nations

Contents

Foreword

This is a foreword written afterward because I believe it is important to share some personal thinking with this book's readers before they begin it.

The reasons I write now, some weeks after completing the last sentence of the amaranth story's last chapter, have to do with credibility — with nothing less important than the truth.

For 25 years as a professional writer and journalist I have developed respect for the truth. More recently a writer of books (this is my fourth) and scores of articles for national magazines, I have worked within the patterns of verity developed over my decades as a journalist.

It was within my professional capacities as a writer that I was contacted by Rodale Press and asked to write a book about the amaranth. That call from Rodale was well timed; at that particular phase of my career, I needed the work. I told the publisher I would do the book.

When I put down the phone, I picked up a dictionary to see what it had to say about the amaranth. Until that day, I had never heard of the plant.

I have learned so much, been through so many experiences and adventures since that day (which now seems so long ago), that another book would be needed to describe the entire saga. I am not going to write that story here, but I do want to tell you enough so you can begin this one with a full understanding of its origins.

My glance at my dictionary was the beginning of an educational process — a research exploration which took me through thousands of pages written about the amaranth, which involved me in literally hundreds of conversations and interviews with individuals who know the plant in each of its many aspects. What I have come to call the amaranth adventure also took me traveling to different parts of this country, and, finally, to a small village in central Mexico: one of the places where the plant has been grown and harvested for thousands of years.

Before my adventure began, I had heard of Rodale Press, was in general agreement with what I supposed to be its editorial posture, but had not been a regular reader of *Prevention, Organic Gardening,* or any of the company's several other widely distributed publications. I and my family — I have since learned — share many of the Rodale opinions; nevertheless, I first met the amaranth and the Rodale organization as a skeptic. "If the plant is as good as they say it is," I asked myself in journalistic fashion, "then why haven't more people heard about it? Why aren't folks like me and my family finding it in our stores and on our dinner tables?"

I still ponder that question, only now from a quite different perspective. Now it is not "if the plant is as good as they say it is. . . ." After my reading, my research, my interviews, my traveling — my amaranth education, if you will — I am convinced of the amaranth's potential. In a way, if you read between the lines of the chapters, you will be reading the story of how my convictions were formed and, finally, set firmly in place. Now I ponder the question of why the crop is absent from the nation's fields not with doubts about the amaranth's intrinsic value as a nourishing food and a plant that is kind to the land, but instead with a full understanding of why any new crop — no matter what its true value — has a difficult time attaining the acceptance needed to make it part of the national diet.

Those difficulties are described in some of the chapters that follow. The personal events which led to my convictions about the amaranth's true significance are missing. That is the reason for this foreword written afterward. I want you, the reader, to understand that over the months and the thousands of miles I learned enough about the plant to know within myself that it is a crop of significance, a plant from our past that our future most certainly needs.

There is considerable reference to Rodale Press, the Organic Gardening and Farming Research Center, to Robert Rodale, Dick Harwood, and other Rodale staff people, scientists, and nutritionists throughout the book. Those references have been a source of some consternation and concern to the Rodale publisher and editors. From both quarters have come suggestions that I eliminate or tone down such references, that this is, after all, a book published under the Rodale Press imprint. "What will the readers think," ask the editors and publisher, "if you write about us? There is always the chance they will think you wrote about us because we asked you to, because we are publishing the book and you are trying to make us look good."

I have, in response to those suggestions, been as restrained as I could responsibly be about the number of references to Rodale Press and its people in the chapters that follow. I have not, however, omitted any piece of information about the amaranth which I considered significant merely because that information somehow related to Rodale and its people.

If there are a good many such references it is because the Rodale organization, more than any other organization, agency, or institution (public or private) is doing more work to study, to improve, and to reestablish the amaranth. That is one of the verities I want this personal foreword to establish.

Which is not to imply that there are not many other individuals, scientists, nutritionists, botanists, geographers, historians, agronomists, institutions, and agencies (public and private) also exploring the amaranth's past, present, and future. There are, and a great many of them are mentioned in the chapters that follow. Just as I am certain of the important roles those cited within have played and are playing, I am equally as certain that I have omitted references to individuals and institutions which should have been included, based on the same criteria applied to their counterparts which were. Such omissions are my oversight, and mine alone. As I have explained, this book has been written from beginning to end as a report on a personal education, an individual effort to start from scratch and to learn as much about amaranth as was possible over those months and miles. There can be no absolute end to such an adventure; I never could have encompassed the total. I gave the effort my best shot, and to those good people and fine institutions in this nation and many others that I have yet to learn about and contact, my apologies — along with the hope that you come forward and identify yourself and your work. This is not the last book to be written about the amaranth in contemporary times, but one of the first.

I am quite certain of that, just as I am convinced of amaranth's potential. And yet if there was a turning point when my growing interest in the plant was converted to absolute conviction, it came not from the pages of an agronomist's report, or a nutritionist's analysis, but from my own observations of life in a small, Mexican village in that nation's state of Morelos.

Some 50 miles south of Mexico City, the village of Tepoztlán is close to the northern border of Morelos — a region known through history for its independence. From the time of the Toltecs, millennia ago, through until Montezuma and the Aztecs, the people of Morelos and its villages resisted encroachments on their traditional ways of life. They did not submit easily to the conquering troops of Cortez when they arrived in 1521, and, some four centuries later when Zapata sounded his revolutionary call, it was the people of Morelos, and, more especially, the citizens of Tepoztlán, who responded first.

They paid a high price for their convictions. Their village was burned, their cattle killed, crops destroyed, women raped, and homes destroyed. Yet Tepoztlán survived and, as it has through history, continued to hold to its traditional and independent ways.

I had learned something of that history before I flew from my home in Maine thousands of miles southwest to Mexico City, and from there by car to Tepoztlán; yet, even if I had not, I think I would have sensed it on the sunny morning I walked into the village square as the people of Tepoztlán made ready for their market day. There was, in the face of every villager, an expression of inner dignity, of remoteness without rudeness, that let me know immediately that I was welcome, but that I would not be patronized; that I could walk anywhere in the village, but that I would never be a part of it.

So I walked and watched as the market was created. To my eyes, the process was a kind of enchantment: the empty square, paved with gray stones smoothed by centuries, filled with people and their wares until, within an hour, what had been neutral space was transformed to a center of color and energy, a visual and human nucleus that vibrated with such vitality, such variety, and such innocent artistry that my heart pounded at the excitement of the event.

Watching what was for the people of Tepoztlán a routine market day, was, for me, a thundering revelation. As an American no different from most of my fellows, I had come to accept the stereotypes of Mexico which are such a pervasive part of our culture. On this, my first visit, every one of the stereotypes was being shattered before my eyes. I was quite unprepared for the experience.

I found the Mexican love of the visually dramatic, and their expressions of that love, on every building, in every market stall, and even in every stitch of their homespun clothing. I was swept away by the colors, by the pure sense of design manifested in a place as basic as a market stall hung with thin strips of beef from a steer that had been slaughtered and dressed that same morning.

To say nothing of the stalls for fruits and vegetables. Banana, orange, lemon, lime, hog plum, papaya, mango, avo-

Author John Cole sampling *alegría,* a popped amaranth grain candy, in the rural Mexican village of Tepoztlán.

cado, squash, tomato, sugar cane, peanuts, coffee, honey, spices, and herbs . . . they arrived in baskets on the backs of women and burros, and in what seemed merely a matter of moments were sculpted, cut, woven, and arranged as works of art. Green bracelets were plaited from garlic, mangoes were slashed with machetes so they became orange flames burning at the corners of an altar of lemons and limes, each set, one atop the other, in pyramids of yellow and green.

There is nothing comparable . . . nothing . . . in these United States. We have neither the visual compulsion to create such instant design, nor do we have the locally harvested produce that would allow such variety and vitality in so small a village.

For the people of Tepoztlán, their care, their pride, their artistry was a heritage of such span that they could construct their market no other way. If they did, centuries would be shattered.

In one corner of the square, a blind man played a flute fashioned from bamboo, and beyond him rose the spires of a church of gray stone outside, a church dancing with colors within.

As my stereotypes fell away, one after the other, I watched as a Mexican man approached on horseback. His horse walked, gently, at the edge of the square. The rider wore white, and on his head was the straw hat of the region's farmers. His machete hung from the saddle; the horse and rider were the epitome of rural Mexico, and their journey had brought them to within a few feet of where I stood.

As he passed me, the rider slowed his mount's peaceful pace just a bit. Looking directly at me with black eyes set in a dark copper face, the rider slowly raised a hand to his hat brim, lifted the hat imperceptibly, and bowed his head in my direction. It was a greeting, a recognition, a "Good morning," to which I had no way of responding in kind. Such artless degrees of courteous dignity are beyond my ken.

It was a small moment, an incident, many would say, of no import, that brief contact with the horse and rider. Yet from it I learned more of the pride, the peace, the palpable dignity of the Mexican people than from any other single incident on my visit.

On the afternoon of that sun-swept day, I began to comprehend some of the reasons for what had become my revelation. I could not comprehend the discrepancies between what I had seen and experienced and the "typically American" long-distance perception of Mexico and its people. But I could begin to understand, as I walked around a mountaintop Aztec temple, something of the centuries which had allowed that Morelos horseman his degree of dignity.

As I stepped from one corner of the market to the other,

rounded a booth, and began to cross the central street of Tepoztlán, I almost bumped into a young Mexican man with the same ancient, Indian features carved on the walls of the Aztec pyramids. He stood before a wooden box, and on top of the box, arranged with typical attention to design, were piled small portions of *alegría,* the Mexican confection made from popped amaranth seeds mixed with sugar syrup and pressed into small cakes.

The man spoke an Indian language, not Spanish, so we could not communicate (my Spanish is wretched and my knowledge of Mexican Indian nonexistent). But looking at his ageless features, remembering how Cortez and his mercenaries had ridden through this very village, burning every field of amaranth, cutting off the hands of the farmers who raised it, remembering the Aztec rituals which utilized the same *alegría* as a religious symbol, and remembering beyond the Aztecs and the Toltecs, back thousands of years to the first recorded utilization of amaranth as a nutritious food by the cave dwellers of Tehuacán . . . remembering all that and the horseman of the morning, I bought and bit into an *alegría* portion, and there in the market of Tepoztlán, face to face with a man from another age, I decided that any crop which had survived as long as amaranth, and owed that survival to a people as independent, as dignified, and as pure as the people around me in that market, was a crop worth whatever it takes to restore it to this nation's dinner tables and the tables of the world.

Readers may find some whimsy, some romanticism, some unscientific reasoning in these details of this pivotal encounter of mine during my amaranth adventure. That, of course, is your prerogative. I said at the outset of this afterward-foreward that it is written with a concern for the verities — that slippery pig called truth that every author and journalist chases every working day.

What occurred in Tepoztlán is the truth as I perceive it. As a person who had to look up the word *amaranth* in a dictionary when I was first asked to write this book, I have spent the months and miles since then becoming fully convinced that our next growing season here in Maine will not go by without amaranth seeds being planted in our own backyard garden.

It is my conviction that each of you, and your neighbors the world around, would be the better for it if you did the same. It is also my conviction that all of us will benefit if and when the amaranth becomes an established crop, taking its gentle place alongside corn, wheat, soy, rice, barley, and the other established grains.

I know how long a step that is — I learned the truth of that during my amaranth adventure. I know how vast is the potential for failure. But I have also learned that failure need not be inevitable. Success is also possible. I am convinced of that. It is the reason I wrote this book. I thought you should know that before you read it.

JNC

Acknowledgments

I am a professional writer, not an agronomist, botanist, nutritionist, geographer, taxonomist, nor any sort of scientist. Yet this book is full of scientific information. It is work taken from others, lifted from research papers, theses, analytical reports, documents which learned individuals spent thousands of hours researching and reporting. I owe them my first debt. It is to them that the primary acknowledgment must go. Without their work, I could have learned little of the amaranth, and written less.

More than any other single individual, it was Nancy Nickum Bailey, a Rodale researcher, who organized and coordinated and filed those works of science and observation. She gathered them, literally, from the world around and made them available to me in an orderly procession. She also checked carefully to see that I did not misuse them, and that proper credit was given. She, more than anyone, was my guide on my amaranth adventure, and, as such, is due our mutual gratitude: yours and mine.

Many others at Rodale Press (and most of them are named somewhere in the chapters) were most helpful. One who is as yet unnamed is Carol Stoner, my editor, who not only did what many editors fail to do—give the book a skilled and thorough going-over—but did it with care and concern for the author. Such editors need more than mere acknowledgment.

ACKNOWLEDGMENTS

Without Bob Rodale, there would have been no amaranth book. But then, he is a modest man; if I took the time to detail his encouragements and his insights, he would see that they somehow vanished from these pages. Thanks are due, however.

And thanks are due to my wife, Jean, who typed, and often retyped, the entire manuscript, making gentle and positive suggestions along the way. She has done the same for each of my books and magazine articles over the years. How I can ever properly acknowledge my debt to her I shall never know.

Amaranth
from the past — for the future

Chapter One

There is a cave not far from the Mexican port city of Veracruz that has sheltered Man for nine thousand years. There are other caves to the north, in the Sierra Madre, and many miles south along the coast of Peru that archeologists and scientists tell us were inhabited nearly ten thousand years ago by Indians whose cultures are still largely a mystery.

But slowly the pieces of the puzzles that are ancient civilizations around the world are being put together, painstakingly, haltingly, and often mistakenly; a new bit of evidence turns up and a piece that was thought to be in place has to be shifted because it will not fit the new patterns of discovery.

The cave in the southern Mexico valley of Tehuacán, in the Gulf Coast state of Veracruz, is known to the scientists who began studying it in the 1960s as the Coxcatlán Cave, and from the crevices and crannies of this deep rock shelter have come remnants of distant history that speak to us across the years in messages able to be translated by the precise instrumentation and analytic techniques of contemporary science.

The careful study of cave residues tells those of us near the two-thousandth year after the birth of Christ that almost eight thousand years before that event, pre-Columbian Indian families had structured a culture, were living in Coxcatlán Cave, and, as later discoveries would reveal, may have been the descendants of yet another civilization that had gone before. There is no certain unraveling of the fragile cultural threads that the centuries have spun. Pyramids, temples, plateaus, Stonehenge, idols, glyphs, carvings, and ancient incantations often lead explorers further into areas they cannot understand, rather than solving mysteries they hoped to comprehend.

But the evidence gathered so carefully in Coxcatlán Cave does tell us, quite irrefutably, some important truths about the men and women and children who lived there. Perhaps most important is the fact of their survival. Culture after culture called the cave its home. In a span difficult to comprehend by those of us who measure human longevity in terms of generations and centuries, the people of southern Mexico lived through earthquakes, fires, floods . . . every type of natural disaster except one: the arrival of invaders from Spain.

The cave was their shelter from the elements; they built fires there to keep themselves warm and to roast what few wild creatures they could capture — and the record shows these were indeed few and occasional. It was not the animals that lived in the hills around them, nor the birds in the trees overhead that insured the survival of those Indians through the tens of centuries. It was the plants that grew in the soil.

Beans, roots, grass seed, millet, cactus, and leaves of several sorts were the basic diet of those Indians of ten thousand years ago. The caves tell us so. Seeds ten thousand years old have been found; sedge roots and cactus remnants — each of them witness to meals eaten by people of cultures like the Ocampo, which occupied the Coxcatlán Cave around 4000 B.C. These were, as has just recently been discovered, primarily a vegetarian people. The nourishment they needed to raise families, to defeat illness, to stay alert and strong enough to withstand natural disaster was acquired from just a few basic vegetable sources.

The question surfaces immediately: by what process of selection did a people like the Ocampo (and the process was being duplicated in every quarter of the globe) choose the plants that would meet their nutritional needs? The answer will probably never be certainly known, but the trial-and-error method appears to be the most likely and is the one that currently has most credibility. Our problem, in these days of instant communication, analytic chemistry, botany, and agronomy, is one of perspective. It is difficult for us to visualize two or three centuries of trial and error as the diet determinant for an entire culture.

Yet, according to most current theory, that is precisely what happened . . . with a little help from the weeds. "Theirs is a long and complicated story," writes botanist Edgar Anderson in his book *Plants, Man and Life,* "a story just now beginning to be unraveled but about which we already know enough to state, without fear of successful contradiction, that the history of weeds is the history of Man."

As they had been for other cultures for centuries, weeds were the friends of the Ocampo. Because some weed species tend to grow better where Man has been than where he has not, some plants came to the early peoples instead of being discovered by them. Ground cleared for firewood outside a cave, or cleared and then converted to a communal dump also quickly became fertile ground for that entire family of weeds which seeks out turned and disturbed ground (and dump sites as well) and flourishes there.

As they were likely to do with every plant that grew within easy walking distance of their home in the rocks, the ancient Indians experimented ever so gently and cautiously with the weeds that sprouted, as if by magic, in the soil just beyond their cave entrance.

A small taste of this leaf, a careful chew on that seed, an examination and a cleansing of the root . . . then a taste of that. And later, perhaps, the grains would be pounded and milled on stone, or heated and popped by the fireside. The classification by this method of a plant first as edible, and second as nutritious, and third (for some plants) as magical may have taken centuries — as difficult as it is for us to grasp in terms of today's temporal rhythms and the pace of change. One aspect of that early system, however, is certain: once established as a food for survival, a plant did not disappear from the culture. The same beans and the same cactus were consumed in Coxcatlán Cave from before 6000 B.C. until after 1500 A.D. when the Spanish conquest wiped out the gentle people of Veracruz.

Until they were skewered on the spears of the Cortez cavalry, the ancient Indian cultures had survived millennia and they had done it through a time when Europe writhed with the torments of the Dark Ages: plague, pestilence, and wars. While the populations of entire cities, and most of those in some nation-states, were being decimated across the Atlantic, the Ocampo and their cultural counterparts through the centuries were maintaining a stable and gently growing culture — a culture that would produce the wonders of the Aztec world — with a basic, vegetarian diet of a selected few primitive plants: all of those, in the beginning, what we would call weeds today, and would be more likely to tear up by the roots and toss in the compost heap than we ever would be likely to try as a staple of our daily diet.

Indeed, it is reasonable to assume that nearly every American with a home garden (flower or vegetable) has, in fact, uprooted and tossed away the direct descendant of one of the six or seven plants that insured the healthy survival of those pre-Columbian Indians for ten thousand years. If you are such a gardener, check your memory and see if you haven't yanked out a pigweed plant sometime recently in your weeding career.

Chances are good that you have. The ubiquitous pigweed (or redroot) seems to be able to find an almost immediate home in open plots of turned soil nearly everywhere in temperate North America. Plucking the thick-stemmed, coarse-leaved plant, with its spiky flower head that turns brown as the summer progresses, from rows of carrots and radishes is second nature for most gardeners, but the next time you do it, pause just a minute to try to comprehend that you have uprooted a plant whose forebears grew at Coxcatlán's doorstep thousands of years ago and helped sustain the life of the people who evolved a culture so sophisticated it could construct the pyramids at Quetzalcoatl, plot the movements of the planets, and, some now theorize, measure cosmic forces.

The plant family, the one that includes both the pigweed and the sustenance of the Ocampo, is the amaranth, the collective label for a plant group that includes more than 60 species that grow on five of the seven continents. The early, grain-crop amaranths that grew at Coxcatlán have been identified by geographer Jonathan D. Sauer as *Amaranthus cruentus* and *A. hypochondriacus,* which still grow in Mexico today where they are now carefully cultivated to produce replicas of the small ivory-colored seeds found in cave crevices where they fell from the hands and baskets of Indians who had learned of their nutritional value six thousand years before the birth of Christ.

Consider what has lately been "rediscovered," if you will, about the food value of the amaranth seed. Those small, ivory seeds not only contain a high-quality starch (carbohydrate) but more protein than other grains and cereals. The two elements together are the basic nutritional building blocks, the food foundation not only for survival, but for growth, alertness, resistance to illness . . . in general, the kind of physical health that also promotes mental energy, creative thought.

What many of us often fail to realize is just how complex a plant seed can be. Perhaps it is the "ordinariness" of the way many of them look: small, indiscriminately shaped, and dull colored. Pull your hand through a goldenrod top in the late autumn, or shake out a bunch of meadow grasses after haying season; there is little in the handful of browning grain that indicates in any way the potential enclosed in each of the tiny shapes.

Not only can seeds regenerate their own species and keep the eternal cycle of growth and reproduction on its natural course, but, properly prepared, many of them can provide the creatures of the planet — human beings included — with the nourishment needed to sustain life and promote growth.

Based on their looks alone, the seeds of nearly every amaranth species are likely to prove among the most surprising when their nutritional potential is expertly and scientifically analyzed. A good deal of that sort of analysis has been done, and the results tell us that the amaranth seed — one of the smallest in size among the grain crops — is one of the largest in terms of the nutritional benefits contained in its chemical and organic package.

Like most cereal grains, the amaranth is a relatively low-fat (unlike some legumes such as soybeans and peanuts) high-carbohydrate seed. This means it stores its energy in a relatively low-calorie system. Yet, if it were solely carbohydrate-fat ratios that were being measured, the amaranth would prove relatively typical.

It is the unique composition of its other basic element — protein — that so distinguishes the undistinguished-appearing amaranth seed from any of its counterparts. To understand why, you need to know a bit about how the human body (a sensitive and complex mechanism if ever there was one) utilizes and absorbs this most important of the nutritional building blocks.

The keys to the delicate system are amino acids. Without becoming overly technical, it can be said that a series of different amino acids are needed to convert consumed protein to proteins the body can utilize. In other words, none of the protein we "eat" goes straight to work creating more nerve cells, building larger muscles, or supplying essential hormones. Before it can perform these and other nutritional wonders, consumed protein must combine with certain amino acids; it is from that internal combination that the body's usable protein will be manufactured.

Protein that is not supplemented with essential amino acids (those the body cannot synthesize) requires the body to supply its own internal additives for conversion. Without the proper amino balance, in other words, the system will "cannibalize" its own protein reserves to absorb those that are entering that delicate and complex digestive mechanism. The net result, of course, is that such consumed protein is sorely lacking in true nutritional value.

One reason meat is more often cited as a protein source is the likelihood that most meats will contain those essential aminos; and one reason many cereal grains are not quite the protein sources they might be is because they lack the critical amino balance.

Most often, that balance missing from most cereal grain amino "mixes" is lysine; in vegetables, the most common missing amino is one of the sulfur-containing acids. If, somehow, the deficiencies could be erased, or chemically supplemented, so that each cereal grain or vegetable could be combined with the perfect balance of the missing aminos — meaning that the protein eaten could be totally absorbed by the body without taking anything from it — the result would be a substance that would score 100 percent on the theoretical charts that nutritional scientists compile.

Such perfection (as in so much of this world) does not exist. Most surprisingly, however — especially in terms of their humble appearance — amaranth seeds come closer to attaining protein perfection than any other grain. On the scientist's scoreboard, those analyzed and tested rate from 75 to 87 on the scale. (Corn, on the other hand, scores 44; whole wheat, 56.9; soybean, 68; and cow's milk, 72.2.)

The reason for the amaranth's unexpected superiority is the particular combination of lysine (one of those "essential" amino acids) and other sulfur-containing aminos that also fit the complex pattern needed to make the entire, subtle nutritional mechanism work most efficiently on the body's behalf.

The amaranth is not perfect. It's not the 100 percent protein the nutritionists set as their standard. Nevertheless, it is surprisingly good at its amino job, and becomes even better if it is combined with more conventional protein sources like meat and wheat. As it turns out, these foods are rich in the few essential aminos that the amaranth does lack; mixed together

— as a hamburger "extender," for example — amaranth flour and meat can move compellingly close to that 100 percent mark on the "perfect protein" scale. And the same results would occur if amaranth were combined with wheat and/or soy flour.

Thus, as it turns out, that ivory-colored, small, modest, and unassuming handful of amaranth seeds that you may one day harvest from your own backyard garden is a protein wonder in its own right, so don't let appearances fool you. They didn't mislead the tens of generations of pre-Cortez Central Americans who, through their own perceptive observations over the decades, learned then what today's scientists have now established: There are few, if any, grain crops as properly protein-rich as the amaranth.

In addition to this most important nutritional asset, grain amaranths are high in calcium (3½ ounces of grain contain more calcium than a glass of milk) and also high in fiber — a substance lacking in many American diets, and one that we are learning (often to our regret) appears essential if we are to avoid many diseases of the colon.

Understanding all this about the amaranth, and being aware of protein shortages around the world, one cannot help but wonder how such a crop could have all but vanished from our tables. If, as we shall see, Cortez had not chosen Veracruz as his invasion port in 1509, the amaranth might well be a part of our standard diet, just as it was for the inhabitants of Coxcatlán Cave thousands of years ago.

Chapter Two

There are many "ifs" in history (If Napoleon had defeated Wellington at Waterloo. . . . If Russia had not sold Alaska. . . .) and they lead to entertaining suppositions. In the case of Hernán Cortez and his conquest of Mexico, however, the "if" factor is reduced to a minimum. Historians can agree: if it had not been for the particular character of Cortez and the grisly circumstances of his invasion, it is almost certain the grain amaranth would now be one of Mexico's staple crops, and would likely be a well-known and popular crop in the United States as well.

Consider the crops developed by past Meso-American cultures which are now a key to the nutrition or industry of many nations. They include: maize, pumpkins, squashes, potatoes, sweet potatoes, tomatoes, lima beans, kidney beans, peppers, pineapples, strawberries, persimmons, peanuts, alligator pears, cassava, quinine, balsam, the guinea pig, turkeys, and rubber. Surely the intelligence and agricultural skills which domesticated, cultivated, and perfected such a host of plant and animal species evolved from a culture of historic significance. And the archeological footprints still clear in the nation of Mexico have confirmed, over the decades, that indeed, there did exist an ancient culture more advanced than most this planet had known.

Cortez was 34 when he landed with his armada of 11 Spanish ships at the Gulf Coast port of Veracruz in April of

1519, but he had already acquired a reputation with his countrymen across the Atlantic as freebooter, good horseman, swordsman, and a great infighter with a knife. He was also renowned as licentious, gluttonous, fond of dice, avid for women and gold. He had come to Mexico (New Spain) to conquer it for his king, Charles V, and, in the process, to make himself rich on whatever he could plunder. To help himself to those ends, Cortez also displayed surprising and often gratuitous brutality toward the vulnerable people he found in southern Mexico.

Ironically, the first place Cortez set foot in the new land was just a short distance from the Coxcatlán Cave, one of the natural homes that sheltered, for hundreds of centuries, the Indian race that had, over those centuries, evolved to found a civilization of culture and complexity. At least a thousand years before Montezuma — Mexico's ruler when Cortez landed — that culture had constructed great cities, built incredible pyramids (the world's largest) and, many archeologists now agree, devised mathematical systems for tracking time, the movement of the planets in their orbits, and, some experts say, recorded their information so it could be understood not only by succeeding generations, but by civilizations in other nations of ancient North and South America.

Many questions remain unanswered: How, for example, were the stones for the great Mexican pyramids moved when the horses that Cortez brought with him across the Atlantic were the first the Mexicans had ever seen? Others are able to be explained with the help of theories that spring from the facts of the pyramids and their inscriptions. It now seems true, as Montezuma told Cortez, that the Aztec empire had been constructed on the ruins of an even greater empire that had preceded it.

It was a common article of faith in Mexico during the sixteenth century that the noblemen who had established that ancient culture, and who had since become gods, would someday return to claim their territory. The omens foretold the date of that return as the year "One-Reed," the very year on the Aztec calendars that Cortez arrived.

The first Indian to sight the armada ran all the way to Mexico's capital, Tenochtitlán, to tell Montezuma that towers and mountains were floating on the sea, bearing strange beings from an unknown world, with light skin, beards, carrying spears, dressed in iron, and riding deers without horns. Montezuma and most Indians were convinced the gods had returned, and they vacillated between fighting to protect their land from being invaded and surrendering to powers greater than any on earth.

The story of the conquest that followed has been told many times, but it is always difficult to comprehend the enormity of the slaughter that did, in fact, destroy an entire civilization. Typical of the ruthless way Cortez dissipated a culture that had taken centuries to evolve was his destruction of Tenochtitlán: the capital city of the Aztecs that once stood where Mexico City stands now.

Writing in *Mysteries of the Mexican Pyramids,* author Peter Tompkins describes a dazzling city:

> *When the Spaniards first glimpsed the island capital of Tenochtitlán — glimmering like an exotic Venice at the end of a wide causeway, with stunning palaces, temples, and pyramids, stuccoed pink with volcanic ash rising from the cerulean waters of the lake — they thought they were dreaming.*
>
> *Though many of them had seen the splendors of Rome and Constantinople, they were amazed at a city of 300 thousand inhabitants fed by fresh-water aqueducts, laced with canals and carefully paved streets, adorned with arcaded squares twice as large as Salamanca's, serviced by a marketplace where 70 thousand Indians daily bought and sold a thousand different products, from filigreed jewelry to mountains of polychrome feathers, including those from the rare quetzal bird.*

> *The Spaniards were enchanted by such luxuries as botanical and zoological gardens, by elegantly towered palaces, higher than the cathedral of Seville, surrounded by large and beautiful private houses with fragrant gardens shaded by bright cotton awnings where courtly nobles feasted on fruits and vegetables, sauces and soups, fish and meat, cakes and pies.*

According to an eyewitness report by Spanish historian Bernard Díaz of the first meeting between Cortez and Montezuma, the Aztec king arrived

> *. . . on a sumptuous litter with a baldachin over his head adorned with light, greenish-blue feathers, gold, pearls, and jade to represent "the verdant blue sky." He was supported by his four principal lords; and the golden soles of his sandals prevented his feet from coming in direct contact with the ground. Other lords preceded him, sweeping the ground and spreading blankets upon it so he should not tread upon the earth. All of these lords did not think of raising their eyes to look at his face — only the four lords, his cousins, who supported him, possessed this privilege.*

It was shortly after that meeting that Cortez and his troops, mean and hungry for gold, massacred without warning the unarmed noble families and warriors of Montezuma's court at a feast-day dance.

A few nights later, Montezuma had his retribution: more than half the Spaniards in Mexico were killed by one of the largest armies of Indian warriors ever assembled. Cortez retreated to the south, assembled a second force, and set in motion a series of events which, within a relatively few years, ended not only the reign of Montezuma, but wiped out for all time a civilization which was surely one of the most advanced on the globe.

CHAPTER TWO

Again, Peter Tompkins:

A year after La Noche Trist, Cortez returned with 450 Spaniards, 40 horses, nine field guns, and ten thousand Tlaxcalan allies. He also brought a secret weapon in the form of a dozen brigantines built in Tlaxcala, carried overland to the edge of Texcoco Lake. To capture the capital city, Cortez planned to advance along three of the causeways on a narrow front with bunched musketeers, cannon, crossbows, and cavalry, using the ships to protect his flanks. However, just before the attack, his Mexican allies, who identified him with Quetzalcoatl, deserted him because the planet Venus had gone into its yearly, eight-day disappearance as the evening star. When nothing bad happened to "Quetzalcoatl-Cortez" during the dark of "Quetzalcoatl-Venus," and the planet reappeared as the morning star, the ten thousand allies flocked back to Cortez's aid. On August 13, 1521, after a 75-day siege, and one of the most remarkable naval operations in history, Tenochtitlán fell to the Spaniards, leaving more than 240 thousand Aztecs dead from wounds or disease. The people of the city, in the account of López de Gomara, one of the eyewitness conquistadores, were

> *. . . tormented by hunger, and many starved to death. There was no fresh water to drink, only stagnant water and the brine of the lake . . . the only food was lizards, corncobs, and the salt grasses. . . . The people ate water lilies, the seeds of colorin, deer hides and even pieces of leather. . . . They ate even dirt. Almost all of the nobility perished; there remained alive only a few lords and little children.*

After plundering the city, Cortez branded the faces of its inhabitants with hot irons, and attempted to force divulgation of the location of the famed Aztec treasure. Though $15 million in gold was found, this represented only part of what was known to exist. Cortez ordered the entire great and beautiful city destroyed; palaces, columns, and gods were buried in the mud. To avoid the stench of the dead, the Spaniards set up camp in the nearby village of Coyacán. Groups of captive Indians razed every last building of the Aztec capital and filled in the canals until not a stone was standing.

For his efforts, Cortez was ennobled by the king of Spain and made marquess of the valley of Oaxaca, captain general, and given a large estate with 23 thousand vassals.

Among the many evidences of an advanced agricultural society destroyed in the razing of Tenochtitlán were the so-called "floating gardens," or *chinampas* as they were known,

and still are, in Mexico. Strikingly similar methods to those used in China — where artificial lakes are created for drainage, fertilized so algae will bloom, and then drained themselves so the algae can be harvested and converted to fertilizer — were employed by the Aztecs. With so much of the land surrounding their capital city in need of drainage ditches, they did not have to construct artificial lakes, but, instead, learned how to make the drainage canals do double duty.

Not only did the canals route the harmless passage of excess water and carry some waste (which acted as an organic fertilizer), but they also provided the base for one of the most efficient forms of agriculture yet devised. Harvesting the algae, water weeds, and other plant material from the canals, the Aztecs mixed it with the rich mud dredged from the canal bottoms, then put the mixture inside frames made of posts interwoven with vines. As the combination of mud and organic matter dried in the sun, it could be scored into grids with hand tools. Each small square, about two inches across, left by the

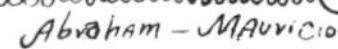

scoring became its own seed pot; as the seeds placed in each square sprouted, took root, and grew, the individual squares were easily separated from the whole and planted in fields outside the city, or in larger *chinampas.*

As a further bonus, the canals were also homes for carp, other edible fish, and a large salamander *(axolotl)* which the Aztecs considered a delicacy.

Before Cortez destroyed them, the *chinampas* of Tenochtitlán had been fertile for hundreds of years; the natural, organic renewal of the gardens through the repeated application of canal mud and algae-rich water — along with the warm climate — enabled them to produce as many as seven different crops a year.

Of those crops, none was more important from both a nutritional and symbolic standpoint than the grain amaranth — *Amaranthus hypochondriacus* — the direct, domesticated descendant of the weed that had taken root outside Coxcatlán Cave and had helped insure the survival of the progenitors of the race that had not only designed and built Tenochtitlán, but had created the rich and mysterious people that had preceded the Aztecs, leaving Mexico's landscape strewn with great castles, temples, towers, pyramids, citadels, and statues so complex, so strange, so charged with secrets that explorers and scientists from all over the world still leave their homelands every year and travel to Mexico in the hopes that somehow they can discover the answer to the riddles left by the ancient Mayans.

Those people had learned, over the centuries, how to select the best amaranth seeds, how to plant them, cultivate them, and care for the seedlings so they became great fields of grain with amazing yields. Sometimes as many as 500 thousand of the ivory-hued amaranth seeds would fall from a single stalk and, along with maize, make up the staple grain of the nation. But, quite unlike maize, the amaranth had characteristics other than high yield and excellent nutritional qualities, and it was those qualities which brought it to the attention of Cortez and the Spanish who succeeded him.

CHAPTER TWO

For reasons that have never been fully explained, the amaranth was assigned magical qualities by the Aztecs — qualities assigned, almost certainly, in Mexico's distant past by the Mayan forebears who surrounded so much of the Mexican Indian's metaphysical life with myth, mystery, plumed serpents, a collection of multiple deities ranging from Chicomecoatl, the goddess of crops, to Tlaloques, the rain god, and a dozen others. Because this staple grain crop was also a key part of the ceremonial pageants which typified a culture he was determined to eradicate, Cortez ordered the mass destruction of the plant, just as he had ordered the total destruction of Tenochtitlán. And, as they had been in the doomed capital, those orders were carried out.

It is, history indicates, often practical, although brutal, for conquerors to not only destroy the armies and navies of the people they defeat, but to also demolish the cultural cement which holds the society in place. That done, the people are totally without defenses, inner or outer, and the conqueror can go busily about the process of exploiting every resource — human and natural — of his new colony. Cortez correctly perceived that the amaranth was more than a mere food staple; it was a ceremonial and religious — perhaps even supernatural — essential. Thus it mattered not that the plant had survived several millennia to become the key crop of the Aztecs; it had a significance beyond carbohydrate and protein and thus, in Cortez's view, had to be absolutely eliminated.

Some concept of the scale of that destruction can be realized from the size of the amaranth tributes demanded each year by Montezuma from the Aztec farmers. It was the king's excessive taxation which contributed to the rural resentments that Cortez converted to allies for his forces. Ironically, the 200 thousand bushels of amaranth grain which had to be brought to Montezuma's palace in Tenochtitlán every year came primarily from the farms of the Indians who joined Cortez when he marched on the capital; yet no sooner had the royal city been leveled, but Cortez and his troops scourged the fields of the Indian's most important crop — *Amaranthus.*

Montezuma demanded much tribute; 200 thousand bushels of maize also had to be delivered, as well as equally staggering amounts of other produce, game, bird feathers, gold, pearls, and clothing fit for a king. The maize, being more bulky than amaranth, actually represented a lesser share of the resource. In absolute terms of what was taken from the fields, the amaranth demands were the highest. In order to meet them, more acres had to be planted with amaranth than any other crop, and it was every one of those tens of thousands of acres that the Cortez cavalry and its Indian allies put to the torch, trampled, cut down, and uprooted.

The sheer expenditure of energy required was monumental, and perhaps would not have been possible had it not been for the amaranth's metaphysical dimensions — the "extra" dimensions that gave the plant an aura of ancient mystery, a spiritual presence that Cortez knew he had to destroy with much the same single-minded frenzy that high priests of the Spanish Inquisition brought to bear against heretics in his native land.

The foundations for the amaranth's "extra dimensions" had been set in place over the centuries by the Mayan molders of that culture's religion, and preserved over the years by ceremonies that had become fixtures of the Aztec calendar. One of the most formally, and strictly, observed of the rituals that dotted the 18-month Aztec year was the one that occurred at a season corresponding to our May. It honored the war god — Huitzilopochtli — and was marked by festivals across the nation and a full-scale ceremony at the temple pyramid in Tenochtitlán. The entire city population attended: priests and nobles in their vivid ceremonial garlands, and iridescent cloaks of bird feathers, golden ornaments, and insignia of rank, stood nearest the temple, surrounded by tens of thousands of citizens, many of them dressed in costume and masks. The ritual was as intricate as the incredibly detailed and convoluted designs traced on the temple walls and the huge, 12-ton, carved obsidian circle in the center of the temple floor that was the master calendar for ten centuries of Aztec time. There were days of dancing — and each dance had its own definition, rhythms, and costume — many processions, chants, and singing, much music of flutes, shell trumpets, and drums. There were also human sacrifices, reportedly greater in number for the Aztec war god than for the other gods honored during the Aztec year.

To add the term *human sacrifice* to the agenda of otherwise predictable activities of a ceremonial ritual is, in these days, a shock to the sensibilities. That there were human sacrifices in pre-Conquest Mexico is not debatable; it is the scale of those bizarre killings that have been (and still are) hotly debated. There are respected scholars who argue that the incidence of human sacrifice by Aztec priests was grossly exaggerated by Spanish historians and in Cortez's reports to his Spanish king. By claiming that ". . . blood-encrusted Aztec priests performed 50 thousand human sacrifices a year. . . ." as Díaz did, the onus was taken off the slaughters of Indians by Cortez and his conquistadores. Instead, they and the entire Spanish presence in Mexico became rescuers of the Aztecs from their own bloody rituals.

Díaz exercised no restraint. "The Indian priests," he wrote, "practice not only human sacrifice, but cannibalism, sodomy, incest, and drunkenness. . . ." and goes on to describe how those priests roasted a man alive, then pulled him out of the fire with grappling hooks and "cut out his still-beating heart." Strong stuff, even for Cortez, who, in later years would report that he had never actually witnessed even one human sacrifice. In France, Voltaire joined the forces of those who believed the Aztecs maligned, if not in the fact of sacrifice, then surely in terms of its scale.

Whatever that scale, the amaranth played a role. The focus of the activity outside the temple pyramid in Tenochtitlán during the war god festival was not human sacrifice, but rather the ritual sacrifice of a huge idol made of amaranth dough and popped amaranth seeds. Temple maidens had begun their work two days before the festival's first day, removing all chaff from the seeds with great care, grinding some of them into a powder, popping others, then mixing the blend with honey and/or syrup. From this doughy material, they constructed a huge idol — a replica of their image of Huitzilopochtli. Decorated with natural coloring — some of it red dye from amaranth flowers — the idol was carried on a litter through the streets of the capital in a great procession which then wound its way through the suburbs and back to the center city. By then, tens of thousands would have joined the parade, following the idol made of the dough-honey mixture the Indians called *zoale* back to the temple. When the ceremony there ended, the idol would be consecrated by the priests as the bones and flesh of the war god, shattered with sticks, and the pieces eaten with "reverence, fear, and tears. . . ." according to Dr. Jonathan D. Sauer.

On that day it was a strictly observed rule in all the land that no other food was to be eaten except the *zoale* of which the idol was made. In return, everyone who participated in the ceremony was obligated to tithe a measure of amaranth seeds.

There were, in Montezuma's time, according to an early chronicle written by a Spanish Franciscan monk named Sahagun, at least seven ceremonies during the Aztec year when *zoale* idols were made, consecrated, and eaten in much the same way they were during the pageant of the war god. Sahagun writes also of a ceremonial feeding of *zoale* "to slaves about to be sacrified. In a final, midnight ceremony, the temple fire was put out and in the darkness each slave was given four mouthfuls."

For the fire god, Xiuhtecutli, *zoale* idols were shaped like birds, carried to the tops of the tallest trees, and, at the close of the festival, thrown down to the ground where celebrants waited to eat the fragments.

There seemed to be no end of the variations on this theme. Figures representing dogs, cats, cows, and snakes were fashioned from the ceremonial amaranth dough. The strange, edible icon had a significance which Cortez could not fathom, but which he properly perceived to be a linchpin of the Aztec culture. In his way of thinking, there could be only one response: the destruction of the crop from which the alien symbols were fashioned. Just as tyrants since have burned books and imprisoned clerics, Cortez surmised he could alter beliefs and national values by eliminating the materials of the national rituals.

Yet, if he had thought a bit more carefully, if he had taken the true sense of the ancient culture he had so traumatically disturbed, Cortez might have realized that his efforts would, in the long run, be in vain. The grain amaranth of the Mexican highlands was not merely a staple foodstuff. It had been, in the tens of centuries that preceded the Conquest, the key to the survival of the Mayan culture, providing carbohydrates when there was no other starch crop, and protein when the ancient Indians of Maya had neither the weapons, techniques, or knowledge to domesticate animals or fowl.

With the certain instincts of a primitive people when it comes to matters of their own survival, those early Indians had come to recognize that their fate and the fate of the amaranth were one and the same. If the amaranth survived, so would the culture. The plant, blessed with the ruggedness and adaptability it had inherited from its weedy progenitors, responded lavishly to the care and husbandry of the first Central American farmers who must have, in those early years, looked on the small, ivory seeds as the key to their very lives.

Thus they spun the threads of immortality woven into the amaranth tapestry, the elements of ritual which honored the gods of their natural world: wind, rain, fire, sun, crops, flowers, feasts, and, when they had grown strong enough for conquest, war. The *zoale* cakes which Cortez sought to eradicate had been kneaded for more than a millennium; they had sprung from a source as vital as life itself, had been made part of the traditions of a culture so sophisticated for its time that there is still no ready explanation for the monuments, pyramids, drawings, and mysteries they have left in the jungles of Mexico.

The amaranth was more than a mere food crop; it was a symbol for a people enchanted by symbols, a people whose wisest leaders had, according to some theorists, established the roundness of the planet two thousand years before Columbus, mapped the globe, computed its size, calculated longitude and latitude, and, some say, traveled to other continents. The mysteries of that travel are as enchanting as the mystery of the lost continent of Atlantis. On the one hand, they can be dismissed as gentle fantasy; on the other, there are facts, realities, artifacts, and presences which can not be explained with any logic unless that travel is acknowledged. In many cases, it is the presence of the amaranth in other nations, other cultures, that proves the mystery's complexity. As reports from five continents establish, the amaranth persists, not only as a plant, but as a symbol of forces we cannot see, influences we have yet to comprehend, spiritual realms we cannot enter.

If Cortez had pondered some of these aspects of the amaranth more throughly, he would have known he could not totally eradicate the plant, no matter how many fields his soldiers burned, or how many Aztecs were killed for possessing *zoale*. What Cortez did accomplish, however, was the effective elimination of the grain amaranth as a major crop — a crop which almost surely would have later crossed the Rio Grande and become a staple of American fields as well as Mexican.

That did not happen. However, neither the amaranth nor *zoale* were erased from the records of Mexican culture. Read this report, filed some 450 years after Cortez, by Daniel K. Early, Ph.D., an anthropology professor at Central Oregon Community College, in Bend, Oregon, who visited Mexico in 1977 to search for remnants of the traditions Cortez had hoped to destroy:

> *By far the main use today in Mexico for amaranth is in making* alegría *candies. To prepare these candies the seeds must first be toasted. This is done on a round clay griddle or* comal. *This* comal *had been used since pre-*

Hispanic times, primarily for cooking corn tortillas, the basic staple of the rural Mexican diet. The comal *is heated over a wood fire or a charcoal* brasero, *a sort of Mexican hibachi. The heat must be just right: if too much heat is applied, the seeds stick; if not enough heat is present, the seeds don't pop. The heat is tested by extending a hand over the* comal *about three centimeters from its surface. The* comal *is ready when a very hot heat is felt from this distance. A small handful, about 1/4 cup of seeds, is tossed on the surface of the* comal *and kept constantly moving with a small whisk broom. The seeds pop up into the air, turning white and doubling in size. With the help*

The first step in making *alegría* is to pop the amaranth seed in a *comal.* A small broom is used to keep the seeds moving so that the popped ones get tossed above the others and do not burn. Note the mat to catch the popped seeds.

The popped seeds are kneaded with sweet syrup made from molasses until the seeds form one large ball.

The sweet mixture is then shaped into a rectangle and cut into squares.

CHAPTER TWO

The *alegría* (to the left and right) is sold as a confection along with other Mexican sweets.

of the broom, the popped seeds come to the surface and do not burn. The unpopped seeds fall closer to the source of the heat. When the seeds all stop popping, they are gathered to the side of the comal. *Then with a flick of the broom, the popped seeds are swept off the side, falling onto a mat on the ground below and to one side of the* brasero.

If the seeds are very dry, they may be moistened with water to promote easier popping. While one person toasts the seeds, other family members are busy preparing the molasses. Cakes of brown sugar, piloncillo, *are broken up and added to the boiling water; roughly 3 kilograms (6.6 pounds) of* piloncillo *sugar are added to 3¼ liters (3½ quarts) of water. Honey or refined sugar can also be used in place of the* piloncillo. *A few pinches of anise are added to the* piloncillo *syrup for flavor, and the mixture is boiled for about an hour. For every 2 liters (2⅛ quarts) of seeds, 1½ kilograms (3.3 pounds) of syrup are added. After an hour or so of boiling, the thickness of the syrup is tested. A stick is dipped into the liquid and pulled out. When a long, thin droplet forms, the syrup is about ready. Then the juice of two limes is added to precipitate coagulation. For the final test, a little of the syrup is placed in a bowl to cool, and a finger is dipped into the molasses and held above the bowl. If the droplet falls from the finger in a long, thin thread without breaking, the syrup has reached its ready point. Another method of testing used is to put some of the syrup between the thumb and first finger and gently separate the two fingers. The liquid should form a long, thin thread. When the syrup has thickened to this point, it is removed from the fire and cooled.*

After the amaranth seed is toasted, it is placed on a sieve, which is made of fly screen nailed to the bottom of a wooden frame about 12 centimeters (about 5 inches) square. The good, popped seeds will not pass through the screen, while the smaller, unpopped or burned seeds pass through, and are later given to the birds. The good seeds are then poured onto a cotton cloth, manta. When all the seeds have been strained, the good seeds are poured into a pile on a straw mat, petate. While one person pours the molasses over the seeds, another mixes them together. The molasses syrup is added until the seed becomes one sticky mass. Kneading allows the air to pass through, and as the syrup cools and thickens, more and more amaranth seeds stick together finally forming one large ball.

This ball of seeds is then spread out into a wooden box, tarima, a mold for shaping the amaranth candies. A metal pipe, somewhat longer than the width of the box, is used as a rolling pin to compress the alegrías into the mold. When compact, the amaranth forms one large rectangle, roughly 40×64 centimeters (about 16×25 inches) by ½ centimeter (about ¼ inch). This large square is then cut into appropriate sizes for sale, using a spatula as a knife and a long, narrow board as a straight edge. The candy makers of Tulyehualco add thin, round, brightly colored wafers of unleavened wheat flour, about 18 centimeters (about 7 inches) in diameter, to the bottom of the frame. These stick to the bottom of the alegría squares and add a little color to attract customers. As the alegría candy further cools down, it becomes firm, and freshly eaten has a crispy, nutty flavor.

The second most common way of preparing amaranth, particularly for home consumption, is to make pinole. First, the seeds are popped on a comal as in making the alegría candies. The popped seeds are then ground into flour on a stone metate. A long, rounded stone, a mano is rolled back and forth over the seeds on the metate, a large stone with a flat grinding surface and three small leg supports. When the flour is finely ground and powdery, a teaspoon or two of sugar is added, according to taste, and the pinole is ready to eat.

Chuales are sweet tamales, prepared only on the most sacred fiesta days: All Souls' Day (the Day of the Dead) and Holy Week. Chuale-making seems to be concentrated in a few villages in the Chiconcuac-Lake Texcoco area. The use of the chuales on only the most holy of days, and the linguistic similarity to zoales, the Aztec amaranth ritual paste, suggests a relationship between the present-day chuales and the zoales used by their Aztec ancestors.

A ball of abejón or aba, Pisum sativum, a type of bean paste, forms the center of the chuale. This is covered with a dough made of either amaranth, blue corn, or huautle flour. Huautle is the name given to the red seeds of huausontle or lamb's-quarters. First the abejón is ground into flour on the metate. Piloncillo sugar and water is heated, cooking into a molasses syrup which is then mixed with the ground abejón and formed into a ball. The amaranth, blue corn, or huautle is then ground into flour, mixed with molasses to sweeten, and patted around the ball of abejón. The tamales are wrapped in corn husks and steamed until done.

The use of the term huautle for the red seeds of huausontle, lamb's-quarters, of the chenopod family is particularly interesting. In pre-Hispanic times, the term huautli referred to both amaranths and chenopods, and there has been some discussion over whether the huautli grain, mentioned in the early chronicles, might have come from a chenopod such as lamb's-quarters as well as amaranth (Sauer 1950:564). Today lamb's-quarters is called huausontle. The same term, huausontle, is also applied to the green seeds of the lamb's-quarters plant. The red seeds of the same plant, however, which unlike the green seeds, are used as a grain, are distinguished by another name: huautle. This differs from the classical Nahuatl term for amaranths and chenopods, huautli, only by a change from i to e of the final i. This seems to suggest *that red seeds* huautle *may derive from the Aztec* tlapalhuautli, *red* huautli, *which according to Sauer, "might have meant either a red-seeded chenopod or a red-leafed amaranth" (Sauer 1950:564).*

This linguistic similarity tends to support the notion that tlapalhuautli *refers at least to a red-seeded chenopod in Michoacan, which was also milled for tamales, although "outside of Michoacan, grain chenopods seem to be unknown in Mexico today" (Sauer 1950:566). Since the red grain chenopod,* huautle, *was found outside of Michoacan in the Lake Texcoco area, well within the boundries of the Aztec empire, the red seeds of the* huausontle *chenopod may well have been used, along with amaranth as a grain crop by the Aztecs during the pre-Conquest period. Samples of the red and green chenopod seeds were brought back for testing.*

Evidence pointing to another relationship between Aztec amaranth terms was also uncovered. Villagers today in the Lake Texcoco area use the term michihuautle *to refer to amaranth,* michihuautli, *in classical Nahuatl. They use the classical Nahuatl general term for amaranths,* huautli, *to refer to mosquito eggs. Sauer writes "*huautli *means nothing but fish eggs or insect eggs to most Mexicans at present. This two-fold meaning apparently based on the roelike appearance of chenopod and amaranth seeds, can be found in the earliest chronicles" (Sauer 1950:567). The villagers around the Lake Texcoco area are said to sometimes add amaranth flour and mosquito eggs,* michihuautle *and* huautli, *to their* chuales, *sweet tamales. If this is true, it seems to suggest that the use of the term* huautli *for amaranth and mosquito eggs may be due not only to their similarity in appearance, but also their combined use in certain ritual foods.*

Were Hernán Cortez to return to Mexico today, he would recognize those "certain ritual foods" in an instant, and would, as well, comprehend the futility of his destruction of the Aztec fields. Had he also known, 450 years ago, just how deeply and distantly in the world of Man the amaranth had taken root, he might never have considered its eradication an even remote possibility. And, in a final irony, he would surely be startled to discover that some Mexican Indians now mold rosary beads from the same *zoale* that was sculpted into idols in Montezuma's time.

Chapter Three

Cortez is dead; the amaranth lives on. Indeed, anyone who gets better acquainted with this surprising plant — even on a casual, nonscientific basis — cannot help but be awed by the strong grip the species has on survival around the world. What prescience did the Greeks have when they coined the words that gave the species its name? Amaranth is derived from two Greek words meaning, "immortal . . . not withering." It could be argued that the derivation is direct, springing from the hardy quality of many amaranth flowers; they last a long time after they are cut and thus are likely candidates to become symbols of immortality.

But, somehow, the amaranth supersedes such conventional conclusions. Incredibly, wherever it has grown, in whatever culture, on whatever continent, in whatever time, it has acquired a further identity and has been seen, in America, India, Africa, China, and Europe, as something more than merely a plant; it has become a symbol, a medicine, an icon, a creation of the earth endowed with a touch of the heavens.

How one plant, which began as a Central American weed, could have acquired such an identity is a story still being told. Even scientists who know most about the plant — people like Jonathan Sauer — acknowledge gaps in their research. There are aspects of the amaranth that go beyond "proof," that belong as much to the poets as they do to botanists and geographers.

AMARANTH, FROM THE PAST—FOR THE FUTURE

Pliny wrote of the amaranth in the Greco-Roman culture; Spenser weaves its image into his long and exotic narrative poem, "The Faerie Queene," and in his epic "Paradise Lost," the great English poet John Milton crowns his angels with "immortal amaranth," an inspiration perhaps for another English poet, Francis Thompson, who included amaranth imagery in his poem "The Hound of Heaven," published in 1893.

A plant of mythology, not merely in the ancient caves of Mexico where it appears in history for the first time, but on other continents as well, the amaranth's adaptability, if not its immortality, saw it make the transition from the early religions of the Mayans to the beginnings of Christianity. The rich purples, reds, and golds of some of the amaranth plant species found their way into monastery gardens of the Middle Ages, moved from there to the altars of cathedrals where the same vivid colors blended harmoniously with the rich and exquisite stained glass windows. Indeed, the globe amaranth is still cultivated in Spain and Portugal as a decoration for rural churches.

Known by a variety of names to English-speaking peoples — love-lies-bleeding, cockscomb, Joseph's coat, and prince's-feather — the ornamental amaranths of England and Europe were also endowed with medicinal qualities. By the seventeenth century in England, some amaranths were considered such essential herbs that the first English colonists carried their seeds with them on voyages to the New World — never knowing, and little suspecting, that for the amaranth that "New World" was an old one indeed.

There was, however, an important difference in the plants American colonists cultivated and those that provided the tribute paid to Montezuma. Old Virginia records tell us that "flower-gentles" (yet another name for amaranths) were planted at Jamestown and Williamsburg and in the walled gardens of many plantations and town houses in Virginia and the Carolinas. The summer gardens of the Governor's Palace at Williamsburg boasted exhibits of pink globe amaranth, carmine cockscomb, and magenta plumes of prince's-feather, waving their varied brilliances above the Queen Anne's lace. John Custis, William Byrd, Jefferson, and Washington grew varied amaranths in their Virginia gardens, later to be dried and used with pearly everlasting and strawflowers in winter bouquets. Arranged in choice glass and china vases filled with sand instead of water, these arrangements graced the winter drawing rooms of many southern homes.

The Pennsylvania Dutch, incomparable gardeners that they were, grew amaranths with pride, and they even flourished in the gardens of Penn's Quakers. Further north, along Cape Cod and in the Bay Colony, the first settlers planted amaranth beside their tansy, yarrow, and teasel. One of the nation's first printed nursery catalogs advertised the amaranth as an "effective plant with plumed tufts of lustrous crimson."

As the nation grew and entered the Victorian era, amaranths were often used as bedding plants in parks and on estates. But their popularity faded with the nineteenth century, and there is evidence (shades of Cortez) of a campaign against them. A Victorian book on flower lore warns that, ". . . love-lies-bleeding draws lightning and should not be planted near a house." And a magazine of the early 1900s calls amaranths those "vulgar, gaudy plants," and dismisses them as useless, weedy, and quite unfashionable.

A plan that drew lightning in the nineteenth century and a plant that incurred the "wrath of the white god . . ." in Mexico in the sixteenth century must have qualities shared by few other things that grow from the ground. Yet, neither the colonists of Williamsburg, nor Victorian arbiters of horticultural tastes were aware of a critical difference between the plants that went in and out of fashion and the plants that had lasted through Mexican millennia.

The plants in Governor's Palace gardens in Virginia had black seeds; the ones in the *chinampas* of Tenochtitlán had white seeds. Black and white: all the colors, no color. There are no more disparate, different values; no gradations easier for the human eye to register. And, in the amaranth's history, the two colors play a crucial role. The black seeds come from plants that are valuable as garden ornaments, as magic, as remedies, as dye sources, as ritual accessories, and, most important, as potherb plants from which young leaves are collected and cooked much the way Americans now prepare spinach.

The white seeds, on the other hand, come from the few amaranth species which have been developed and domesticated as a source of cereal grain — a grain high in protein and carbohydrate, a grain which, by itself, can supply nutritional essentials.

Scientists who have made a study of the plant are now convinced that the two species — the white-seeded, grain-bearing plant and the black-seeded potherb ornamental — do not crossbreed with each other. For reasons which may well be locked in caves deeper than Coxcatlán, the white-seeded plants, domesticated over the centuries from the first, wild amaranth, will remain white-seeded, and must be planted by Man to continue to survive as a food crop. The black-seeded plants, on the other hand, tend to be able to shift for themselves, to survive, as do all plants of the wild, with the miracle of their own regeneration.

What must have originally begun as a black-seeded plant has, through the Mayan millennia, been transformed to a white-seeded one. What magic touched it has not yet been discovered; what good fortune preserved it through the tens of centuries cannot even be guessed; and what forces transported it around the globe are still a mystery of mysteries wrapped in the black and white symbols of darkness and light, of evil and good, of vice and virtue, of death and life . . . universals as ancient as time itself, symbolized by a humble plant that sprouted along riverbanks eons before Man claimed the first cave as his home.

If white and black are the colors of life and death, then red is the color of violence, the color of blood, the color of cardinals, the devil, fire, and danger. And red is also the color that comes from the amaranth. For a plant whose history is interwoven with images of the metaphysical and supernatural, the colors black, white, and red are not only the most fitting, they are the most basic, the most symbolic in every culture, on every continent, in every corner of the world. Like those colors, the plant itself has established a global presence, just how, no one is yet certain. Of those who have tried to trace the amaranth's journey from the beginning of time, few have been as diligent, persistent, or have traveled as far as Jonathan Sauer whose work on the taxonomy and ethnobotany of grain amaranths began back in the late 1940s. Among his many reports, writings, and papers, is the following brief essay on the plant's global history:

> *When prehistoric people began exploiting wild amaranths, they probably did them more good than harm. Whatever the species lost by being eaten must have been more than made up by new opportunities for them to grow in places disturbed by human action. That is because wild amaranths, like the ancestors of all our important food crops, are ecological pioneers. There is no place for them in forests, grasslands, or other dense vegetation. Until people began disturbing such vegetation, amaranths must have been rather rare plants that were confined to naturally open places like riverbanks, desert washes, lakeshores, coastal marshes, sea beaches, and dunes. Consider for example* Amaranthus powellii, *the probable wild ancestor of* A. hypochondriacus, *the domesticated grain crop. Major John Wesley Powell, after whom it is named, collected its seeds in Arizona on one of his Grand Canyon exploring expeditions. It was native to the canyons and arroyos of the western United States and Mexico. It commonly grows on fresh sand deposited by flooding streams.*
>
> *Like some other vigorous pioneer plants, amaranths are especially efficient in hot, bright sunshine. They are poor competitors when crowded or shaded but are tough*

and quick-growing in the open. Seeds of the wild species are scattered widely by birds, flowing water, and other means and may lie dormant in the soil for years until some disturbance opens up a place for them to grow.

Such plants were superbly adapted to become weedy camp followers of ancient hunters and gatherers. They would have been quick and prolific weedy colonists of village garbage dumps, trash heaps, and shell mounds. There they would have had plenty of nitrogenous manure. So, without anybody planning it or even thinking much about it, a welcome new source of conveniently available food was placed on the doorstep by these volunteer weeds. This symbiosis between people and useful weeds was the first step in crop domestication. Later, when human populations built up to such numbers that people had to abandon their old, easy-going, hunting-gathering life and become hard-working farmers, they inevitably turned to such useful, familiar weeds for their first cultivated crops.

The domesticated amaranths are no longer just tamed weeds. Selection of thousands of generations of the crops by hundreds of generations of farmers has changed the plants into creatures as different from their wild ancestors as a dachshund or a collie is from a wolf. The purposes for which amaranths were selected were quite different in Eurasia and in the Americas.

In Eurasia, amaranths were domesticated mainly as leafy vegetables, as were their relatives, spinach and chard. A. lividus, *a low-growing, succulent, purplish red herb, was cultivated in ancient Greece and Rome and in medieval European gardens but it has now been almost completely replaced by other green vegetables that people like better. By contrast, several Asiatic amaranths, such as* A. tricolor *and* A. oleraceus, *are popular green vegetables that might be worth trying in your own gar-*

CHAPTER THREE

A. tricolor, a popular Chinese vegetable amaranth prized for its tender green leaves that taste much like spinach.

den. In many regions of India, China, Japan, the East Indies, and the South Pacific islands, these amaranths are the favorite leafy vegetables. They have been in cultivation for over two thousand years and many different cultivars have been developed, some with brightly colored leaves. One of these is sold in flower seed packets under the name of Joseph's coat.

American Indian farmers can be credited with domestication of all the grain amaranths. In 1492, grain amaranths were important crops in two large regions. One extended through the Andean highlands of Peru, south past Lake Titicaca in Bolivia to the temperate valleys of northwest Argentina. The other extended from the highlands of Guatemala, northwestward through Mexico into Arizona.

The grain amaranths were grown by many different tribes speaking many languages and calling the plants by many names. Usually only pale amaranth seeds were planted for grain. Plants grown from the dark, wild-type seeds were used mainly for greens. Pale seeds in amaranths, like white fur on a mouse or white feathers on a duck, are due to genetic mutations selected by the domesticators. The pale seed mutation must occur extremely rarely, since it has never been reported historically in wild or weed amaranths. When it occurred, some observant ancient Indians must have carefully saved and multiplied the new kind of seed. The pale seed character seems to be connected with better popping quality and better flavor of the grain. It may also have something to do with prompt seed germination. Given moisture, the pale seed comes up promptly when the farmer plants it rather than breaking dormancy on its own schedule like the wild type.

Having a pale-seeded crop was exceedingly important to the Indian farmers in another way. It gave them a label for the crop that helped to keep it genetically pure, even though wild and weed amaranths grew near enough for cross-pollination. Careful Indian farmers eliminated dark seed from the grain saved for sowing and in so doing eliminated many mongrel hybrids. This made it easier to improve the crop by selection for greater yield, color, and whatever other characteristics the Indian farmers cared about. I am not suggesting that prehistoric farmers deliberately planned to improve the crop. Farmers and gardeners are plant breeders whether they set out to be or not. The mere fact that they select some plants for reproduction and reject others causes evolutionary change regardless of any awareness of genetic theory.

You can see how successful prehistoric breeders were if you compare a wild A. powellii *with a domesticated* A. hypochondriacus. *The wild species has a fairly respectable seed yield, a single healthy plant maturing thousands of seeds, but the domesticate escalates this to hundreds of thousands of seeds for every one planted. The ratio of the amount harvested to the amount sown is far greater than in any commercial grain crop. The prehistoric breeders also had spectacular success in breeding brightly colored grain amaranths. The red pigment, chemically similar to that in beets, is probably neutral nutritionally and was selected for purely aesthetic and mystical value. It must have taken long, careful selection to develop the gaudy domesticates from the dull-colored wild species. Many people see a field of grain amaranths as the most beautiful of all crops.*

There are three distinct species of grain amaranths and we will sketch their histories separately.

Amaranthus caudatus

An ornamental form of this species with pendulous, blazing red inflorescences is commonly sold in seed pack-

A stand of *A. caudatus,* noted for its pendulous flowers.

ets as love-lies-bleeding or red-hot cattail, a name shared with unrelated plants. Other forms of the species give much better grain yields. One good grain variety that has club-shaped inflorescences is sometimes classified as A. edulis, *but I believe it is just a mutant form of* A. caudatus. *This crop originated in the same region as the common potato, the Andean highlands. The Spanish conquerors first encountered it in the Inca empire and called it Inca wheat, but it is much more ancient than the Incas. It was domesticated more than two thousand years ago, for some of its pale seeds that were placed in tombs, evidently as food for the dead, have been shown to be that old by radiocarbon dating. The crop is still widely grown in the Andean region, mostly by conservative Indian peoples. It is usually planted in small patches close to houses, not in large fields as a staple crop. The amaranth grain is toasted and popped, ground into flour, or boiled for gruel.*

It is considered especially good for children and invalids. The crop still contains a great deal of genetic diversity in South America, only a small sample of which has been introduced to other continents.

Evidently the Spanish took seeds back to Europe promptly and it soon spread through European gardens as an ornamental, as shown by sixteenth- and seventeenth-century herbal books. Around 1700, it was sometimes grown as a grain crop in Central Europe and Russia, being eaten as mush and groats. By the early nineteenth century, it had been taken to Africa and Asia, where it is now planted as a grain crop in such widely scattered regions as the mountains of Ethiopia, the hills of South India, the Nepal Himalaya, and the plains of Mongolia. In Asia, most of the people who grow this species also plant larger quantities of A. hypochondriacus.

Amaranthus cruentus

This is one of the most ancient crops domesticated in America. In the famous Tehuacán caves in an arid part of Mexico, archaeologists have dug up some of the oldest dated remains of New World crops. In the caves, remains of A. cruentus, *both the pale grain and bundles of plants brought in for threshing, have been found at a dozen levels, dating from 55 hundred to a thousand years ago. The species is still grown in the region and popped amaranth seed cakes are sold on the streets of the modern town of Tehuacán.* A. cruentus *has also survived, at least until recently, as a grain crop of a few Indian villages of southernmost Mexico and Guatemala, but it is a declining relic. Thirty years ago, I found it as a minor but regularly planted grain in various Maya villages of highland Guatemala where nobody plants it any more. It was being abandoned along with other Indian traditions even before those particular villages were devastated and demoralized by the great 1976 earthquake. None of my old collections of pale* A. cruentus *grain are still viable and I don't know where you could get viable seed. It would be a shame if this crop were to become extinct before its potential can be evaluated.*

The dark-seeded forms of the species are doing much better. A very deep red kind, sometimes known as blood amaranth, is often sold as an ornamental in commercial seed packets.

During the nineteenth century, this deep red A. cruentus *was adopted as an ornamental and for cooked greens by gardeners all around the tropics. It is grown in many places in Asia, the East Indies, and the South Pacific. It became a more important crop in tropical Africa than anywhere else.* A. cruentus, *like corn, sweet potatoes, and other American Indian crops, was evidently introduced to Africa by Europeans and was then passed quickly from tribe to tribe. It outran European exploration of the interior so that Livingstone and many other mid-nineteenth-century plant collectors found it already in cultivation when they first arrived. A hundred years ago, the species was reported to be a favorite source of red dye in the Lake District of East Africa. Today it is cultivated in much of Africa as a green vegetable, being planted and gathered all year round in the humid regions. In parts of West Africa, it is the most important of all green vegetables. The tender young seedlings, cultivated intensively on commercial truck farms, are pulled up by the roots and sold in market towns by the thousands of tons every year.*

Amaranthus hypochondriacus

This is the most robust and the heaviest-yielding grain amaranth. It was probably domesticated in central

A. hypochondriacus

Mexico, farther north and later than A. cruentus. *It first showed up in the Tehuacán caves about fifteen hundred years ago but by then it was fully domesticated, with pale seeds, and it may have been cultivated elsewhere for a long time before. Like corn and beans,* A. hypochondriacus *was grown by Arizona cliff dwellers in prehistoric times.*

The same amaranth grain continued to be an important food in Mexico after losing its role in pagan rituals. A royal inquiry in 1577, which asked for information on crops of each municipality, revealed that amaranth grain was still important in much of its old area. During the next three hundred years, its importance declined during the disintegration of Indian cultures. By the 1890s another agricultural census showed that grain amaranths were no longer a major crop in any large region of Mexico. Thirty years ago, when I went hunting for relics of their cultivation in Mexico, it was not hard to find small patches of them planted here and there through their old area, but not by many people and not in every pueblo. A similar decline took place sooner in the southwestern United States. It has been about one hundred years since the last record of Indian cultivation of the crop there.

A. hypochondriacus *was introduced to Europe before 1600. Both pale- and dark-seeded forms were originally grown there but European gardeners were only interested in the species as an ornamental and did not continue to select pale-seeded grain forms. The dark-seeded ornamental, commonly sold in seed packets as prince's-feather soon took over.*

Meanwhile, the pale-seeded form was introduced to Asia, perhaps by way of Europe by the Dutch colonists in Ceylon, where it was being cultivated by 1800. By 1850, it had become established as a regular food grain among the primitive hill tribes of South India and also in remote parts of the Himalaya. Before 1900, it was being cultivated from Persia and Afghanistan through northern Kashmir and Nepal and the interior of China to Manchuria and eastern Siberia. Reminiscences and diaries of travelers in Asia during the nineteenth century have much to say about grain amaranths. The rich, red plants glowing in the sun caught the eye, often from far away on a mountain trail. The crop also attracted attention by its success at high elevations with very short growing seasons. It was planted as high as buckwheat or any other crop, for example being grown above nine thousand feet on the Tibetan frontier. In the Himalaya and other parts of Asia, the grain came to be used much as in Mexico. It is popped for sweet confections, milled, and made into gruel and tortillalike bread.

European travelers who saw this crop being grown by isolated people in remote regions assumed that it was an ancient native of Asia. We now know this is out of the question because no wild Asiatic amaranths are closely enough related to A. hypochondriacus *to have been its progenitor. The Asiatic crop is obviously a narrow, genetically uniform sample of the more diverse Mexican crop. I once suspected the crop might have been introduced to Asia by prehistoric voyagers, but this now seems unlikely. There are many other American Indian crops, including corn, chili peppers, tobacco, and peanuts that were probably brought to Asia by Europeans since Columbus. These were taken over so quickly and enthusiastically by Asiatic farmers that they outran European exploration of the back country. These farmers knew a good thing when they saw it.*

In India, A. hypochondriacus *has continued its rapid spread into the present, coming out of the hills and mountains onto the plains. It is commonly planted both in small garden patches and in large fields as solid stands. Hundreds of thousands of pounds are shipped to Delhi and other large cities. Among Hindus, popped amaranth grain soaked in milk is the only food permitted on certain festival days. The Indian government's Council of Agricultural Research is promoting expansion of the grain amaranth crop, and geneticists in India have begun the first program of scientific breeding. Until recently, India was the only country where the potential of grain amaranths was generally appreciated.*

There seems to be only one record of dye amaranths in the Old World: in East Africa an amaranth species native to South America is cultivated for a much-like-red dyestuff. Recorded occurrences in the New World are more numerous, but widely scattered. In Bolivia and Argentina an amaranth is grown for its inflorescences, which are used to color alcoholic beverages. In Ecuador, Indians use the plant to make a dye for their ceremonial offerings to the dead on All Souls' Day.

The indicated geographic distribution of the dye amaranths shows some resemblance to that of the grain amaranths, but the coincidence of the two is far from complete. Dye use appears to be a distinct culture trait, more localized than grain use, yet present among some peoples who do not grow amaranth as a grain crop. The full story of the origin and spread of this trait and of the plant species involved cannot be worked out until more adequate data are collected.

We do know that since the 1890s a cultivated amaranth, called komo, *has been reported occasionally among the Hopi Indians of the southwest. Some of these reports suggest the plant was introduced, but do not specify how long ago, or from where it came. They all agree that it is grown in irrigated terrace gardens and is used to give a red color to ceremonial* piki *or wafer bread. In 1948 the Hopi artist, Fred Kabotie, wrote me concerning* komo*:*

> *The Hopi people, as far as my memory stretches, have been raising this plant on their irrigated, terrace gardens. The stalks and leaves are dark red. The Hopis use the flowers*

> *and seeds for coloring their* piki *bread during Katchina dances. The flowers and seeds are soaked in water overnight. The following day the juice is strained out through a yucca tray or cloth. This coloring matter is used in place of clear water in mixing a fine, white cornmeal into a thin batter in a bowl. The women cook it over a fire on a hot, flat stone. . . . These red* piki *breads are made into rolls that are given to the children by the beloved friends, the Katchinas. . . .*

> *Amaranths cultivated in little gardens by the Zuni Indians were ground to a fine meal and used to color the maize wafer bread which is carried by impersonators of gods and thrown to the people between dances.*

In general, a tidy, brief history, wouldn't you agree? With the exception of one or two places, like the lines that read: "I once suspected the crop might have been introduced to Asia by prehistoric voyagers, but this now seems unlikely. . . ." Jonathan Sauer is a model of scientific certainty. When we finish his compact, well-written, and quite detailed story of the amaranth's past, we are tempted to believe we know just about all there is to know on the subject. If we are questioned about the plant, all we need do is turn to these pages and there will be the answers.

It is a case of being comforted by what seems certain, because less is known about the amaranth than would appear, even if we try reading between the lines of this reassuring history. What we have to do is probe further, to look at another, longer, more detailed paper on the plant, also written by Jonathan Sauer. When we finish this, we are back essentially where we began: with more questions raised than answered.

Consider, if you will, the following lines from a paper on the amaranth written by Dr. Sauer for publication in the *An-*

Organic Gardening editor Jack Ruttle (left) with Dr. Jonathan Sauer (right), who is noted for his outstanding work on taxonomy and ethnobotany of grain amaranths.

nals of the Missouri Botanical Garden. Taken in the collective sum of their parts, they do almost more to create a mystery than to resolve one. Try following the patterns that emerge as we go through the manuscript and pluck out the quizzical threads along the way:

> *The role of grain amaranths among the Indians of the United States is an unsolved puzzle. . . . The trail begins very faintly with only a few hints and fragments of information. . . .*
>
> *There are some puzzling historical records which report grain amaranths in the Southwest. . . . The only information indicates that the seeds were used by the Arizona Indians for food. The seeds are light colored, and therefore almost certainly from a cultivated plant. I know of no example of a wild or weed amaranth with pale seeds.*
>
> *Some writers made cryptic statements that the Arizonans cultivated the plant, but gave no details. . . .*
>
> *At the present time, the Mexicans cultivate some chenopods which look so much like grain amaranths that the two are often confused by casual observers. The story of these two plants cannot be traced very far without leaving the firm ground of certain identification.*
>
> *Many of the modern records and practically all of the older accounts are written by persons who knew little of botany. . . . The best clues in the early literature are the crude drawings in two sixteenth-century writings . . . the two plants have probably been so confounded that their stories may never be disentangled. . . .*
>
> *Shortly before 1900, Lumholtz visited the remote Huichol (Mexico) village of Haiokalita in the mountains of northern Jalisco and watched these primitive Indians celebrate their greatest feast of the year. Although he did not realize it, Lumholtz witnessed a ritual which was strangely like those practiced far off in central Mexico more than three hundred years before:*
>
> > *In front of the drum was spread a blanket, and on this a great number of cakes were de-*

> *posited, shaped to represent animals, such as the deer, turkey, rabbit, etc. They are made from a certain seed which is ground and mixed with water. The seed and the plant from which it is obtained, are called* wa-ve; *being yellow, it belongs to the god of fire. The tribe probably made use of this grain before it had corn. At present, it is used chiefly for ceremonial purposes, except when the corn runs short. No one is permitted to eat it until the race is over.*

The festival culminated in a ceremonial race for life. As each runner finished, he pierced a cake with a straw and handed it to one of the elders, who, in eating it, "secured health and life for all." Wa-ve *was cultivated on a small scale and Lumholtz recognized it as a species of amaranth. . . .*

In the late nineteenth century, Edward Palmer visited an Indian village near Guadalajara and saw rosaries made of little balls of amaranth dough, for something called the Feast of the Cat on December 8. The main, modern use of the grain amaranth in this region is for little cakes of popped seed bound with syrup . . . for some reason, they become particularly abundant and conspicuous on special saints' days. . . .

If there are any ancient records of grain amaranths in Guatemala, I have been unable to find them. Even in recent literature, the only published reference to the plants seems to be in the single statement that around Coban the seeds of the amaranth are mixed with brown sugar to make a confection. The crop is practically unknown to everyone except the Indians who grow it. After traveling halfway across the country and questioning not only Indian farmers, but also professional botanists and ethnologists, I had drawn a complete blank and was convinced there were no such plants in Guatemala. Finally, the ethnologist Rodas, a Quiche Indian himself, told me about cultivation at San Martín Jilotepeque. On the way there, the bus driver, a part-time farmer who had driven the route for years, assured me that nobody grew any such plants in that area. When I pointed out some of the big, red, gaudy plants in the maize fields along the road, he stopped the bus to get out to look at them and was astonished when the local people confirmed my strange notion that the seeds were good to eat. . . .

In contrast to Mexico, the historical record of the grain amaranths in the Andes is extremely poor. . . . [yet] the tremendous ceremonial role played by the plant in the Aztec empire was taken over by maize in the Inca empire. The Inca Feast of the Sun is startlingly like the Aztec Feast of Huitzilopochtli. It, too, was the greatest feast of the year, held in the capital city with bloody sacrifices and colorful processions of all the chiefs and dignitaries. Just as in Mexico, the temple maidens on the night before the feast prepared a sacred bread to be distributed and eaten ceremonially. . . .

Thanks to a Jesuit chronicler, there is a record of grain amaranths in Peru. He wrote that red and white bledos were a very common food of the Indians, and in the city of Guamanga very pleasing candies were made of the white seeds mixed with sugar. . . .

Until the present century, no further notice of the crop seems to have been taken by anyone, except the Indians. . . .

Herrera believed the plant was an important, ancient, possibly pre-Inca crop of the temperate valleys of Peru, which has declined to a relic in recent times. Pulgar Vidal reported that it was an ancient widespread crop of the highlands, "but is unknown in city markets, its consumption being limited to the small harvests that each person obtains from the fringes of the maize fields. . . ." Other persons who have seen the crop in Peru agree that it is grown only by the Indians in small plantings. . . .

In an archeologic site, believed to be pre-Columbian, at Pampa Grande in the Argentine province of Salta, a funerary urn full of seeds was recently unearthed. Mixed with maize, bean, and chenopod seeds were a quantity of amaranth flowers and pale seeds. Since this species is unknown in the wild state, this discovery is good evidence that the crop is ancient in northwestern Argentine. . . .

We shall see that the Asiatic and American grain amaranths are very closely related and that there is reason to believe that the entire group must have been originally domesticated in America. The question of how long they have grown in the Orient is therefore of considerable interest.

The crop is scattered so widely through Asia and is so firmly entrenched among remote peoples that it gives a powerful impression of great antiquity in the area. Many observers have concluded that the crop has certainly been cultivated in southern Asia from time immemorial and probably originated there.

The ancient Chinese botanical sources have been carefully examined. It is hardly to be expected that grain amaranths would be recorded in these sources, since one can search long and hard and vainly for any mention of them in modern literature on China. Only one definite account of the crop in modern botanical literature in modern China seems to have been published. Therefore, it is startling to find that Bretschneider presents what seems to be a clear reference to a grain amaranth in an ancient Chinese materia medica. This work, written about 950 A.D. for the prince of Shu — modern Szechwan — lists six kinds of hieu, *a generic name for a group of related plants, mostly indigenous amaranths cultivated as potherbs. Among these were two whose seeds were used in medicine; one was called* jen *(man)* hieu, *possibly because it grew tall and erect. . . .*

Such cryptic records are of little value . . . although they suggest that the grain crop may have been grown much more widely in China than the more definite records can prove. . . .

In South India and Ceylon, Buchanan-Hamilton named the white-seeded south Indian grain amaranth A. frumentaceus. *He discovered it "on the hills between the Mysore and Coimbatore countries where the natives call it* kiery *and cultivate it for the seed which they convert to flour and which forms a great part of their subsistence. . . ."*

The crop has persisted in these hills into the present century. It has been reported that hillmen on the Anaimalai Hills cultivate an amaranth at elevations of two thousand to four thousand feet.

In northern India, Buchanan-Hamilton gave the name A. Anardana *to the white-seeded plants he found cultivated in the fields of Nepal, Bhagalpur, and Cosala. In his notes, he also mentioned that one of the chief crops high in the hills of Kullu was a plant called* chuya. *Madden believed that* chuya *and* anardana *were one and the same thing, and that* anardana *was an incorrect rendering of* amardana, *as Buchanan-Hamilton once wrote it, meaning "immortal grain." He had seen this tall, crimson or yellow amaranth cultivated all over the Himalaya, under various local names, including* ramdana, *grain of the god Rama. The names given this crop in different regions of the Himalaya are bewilderingly variable. . . .*

Later records show that grain amaranths are scattered widely across northern India and the neighboring states. Little precise information is available on the distribution of the crop in the lowlands. . . . Behind the plains, in the foothills and mountains of the Himalaya system, amaranths have become an important field crop and a conspicuous feature of the landscape. They have been reported repeatedly along the whole length of the Hima-

laya from Kashmir to Bhutan. Important early collections of grain amaranth were made by the Schlagintweit Tibetan expedition of 1856. . . .

Pant gives a fascinating account of the precarious agriculture practiced by the Mongoloid nomads around Milam on the Tibetan frontier of Kumaon. The area is too high for permanent habitation. It is covered with 6 to 12 feet of snow for more than half the year and the growing season is only about four months long. Crops are raised by primitive hoe cultivation in tiny, steep plots studded with glacial boulders; the boulders aid the growth of plants by storing the sun's heat. The crops are always stunted, often ruined by persistent mists or avalanches; sometimes they are buried by early snows before they can be harvested. In spite of these tribulations, the people sow, and usually reap, barley, buckwheat, and grain amaranths, sometimes at elevations of more than 3,500 meters (13,734 feet).

The Himalayan plants, like the Latin-American ones, are rich and variable in color; this striking coloration is partially responsible for what little attention they have received in the available literature. Madden was enthusiastic about the brilliant crimson and rich yellow of the plants: "The effect of a mountainside, terrace above terrace, covered with distinct fields of these colors, and glowing under the rays of the afternoon sun is gorgeous indeed. . . ."

Many of the writers give more space to the aesthetics than to the uses of the crop, particularly since the grain does not enter into commerce. Visitors to the hills of India are inclined to smile at people who live very largely on these minute seeds; yet Watt wrote that "This grain is one of the most important sources of food to the hill tribes of India. . . ."

In China, the seeds are know as tien-shu-tze, *or millet from heaven; they are popped and mixed with hot sugar syrup made from fermented rice or millet grain. When the mixture cools, it is cut into thin slices; the candy is crisp, puffy, sweet, and well flavored. On the Chinese New Year's Eve, in every farmhouse, quantities of such candy are made.*

The crop is probably also present in the southwest of Szechwan in Yunnan; amaranths are reported extensively cultivated on the upper Mekon above Weisi, and also around Tali. Although no mention of use is made, it is difficult to imagine such large-scale cultivation as anything but a grain crop. It is conceivable that the crop may extend almost continuously through these poorly known parts of interior Asia from Szechwan to the Himalaya.

On the other side of Szechwan, the trail is not much clearer. Pale-seeded forms of species ordinarily grown for grain have been collected in cultivation in Kweichow and Hupeh, but with no data on use. Small-scale cultivation of amaranths for their edible seeds has been noted in Honan and Hopeh, but no details are available. . . . Yet, in northern Manchuria, there is a definite account of grain amaranth cultivation. . . .

In Africa, Baker and Clarke (1909) claim that amaranths are cultivated throughout the warm continent as a grain. This statement may be an exaggeration based partly on simple records of the occurrence of grain amaranth species. In Africa, just as in Asia, some of these species have been widely diffused as garden ornamentals; occasionally they are also cultivated as potherbs. I am unable to find any confirmation of grain use of the plants in Africa, except for Dalziel's statement that in Sierra Leone an introduced amaranth is said to be cultivated for the seed as food. . . .

As for the intriguing problem of when and how the species were introduced to the Old World, the evidence remains inconclusive. There is no proof that any of them were in the Old World before Columbus. On the other

> *hand, their spread to Asiatic and African cultivation certainly outran European contacts and could scarcely have been initiated by any direct transfer between Old and New World colonies of the same European power. Introduction through Europe itself seems most likely in the case of the Andean* A. caudatus. *Pale-seeded forms evidently were taken there directly from South America in the sixteenth century, long before they were found in Asia or Africa. However, dark-seeded forms that are minor components of the Andean crop somehow became established in remote parts of Asia and Africa before they were known in Europe.*
>
> *The Mexican* A. hypochondriacus *and the Central American* A. cruentus *evidently reached Europe in the eighteenth century, both arriving via Asia and directly from America. Europe got only dark-seeded forms that are inferior as grain and grown mainly as potherbs and ornamentals. The pale-seeded forms of* A. cruentus *apparently never left Central America, but pale-seeded* A. hypochondriacus *had reached Asia by the eighteenth century. At present, India alone undoubtedly harvests more grain from this one species than is produced by all the relics of grain amaranth cultivation in the New World.*

Thus does Jonathan Sauer document the history of the amaranth and its worldwide travels. Another authority, however, has a diametrically different view. Writing in a paper prepared for the Indian Council of Agricultural Research in 1961, Harbhajan Singh, a plant introduction officer of the Indian Agricultural Research Institute in Delhi, argues that Asia is the amaranth homeland:

> *Dr. Sauer has given us an interesting and informative account of the early history of grain amaranths on four continents. His observations are based on a careful*

study of various writings available on the subject. According to him, the conclusion that grain amaranths are all of New World origin seems inescapable. Abundant direct evidence of the antiquity of the crop is available from the New World. Early documentary records from Mexico indicate that the grain amaranths were one of the great food staples of Mexico at the time of the Conquest, and regarded by Mexicans as among their most ancient crops.

Many other investigators, however, have concluded that grain amaranths have been cultivated in southern Asia from time immemorial and were probably originated there. Merrill suggested (1954) that these were introduced from Brazil to India (Malabar coast) by early Portuguese traders, after 1500. From the south, it later reached northern India. There is, however, not much direct evidence available on the early story of grain amaranths in northern Asia. One important reason for this inadequacy of information may be the concentration of this food grain in the inaccessible highlands less well known to the ancient Oriental writers and early European travelers. Sauer has suggested that a study of the old Sanskrit writings in India, where the crop is more widespread, is important. . . .

Amaranth grain has been reported to be more nutritive than the common and prized food grains. The main virtue of the seed lies in its high protein content coupled with an easily digestible carbohydrate component. The growth-promoting value of the seeds is reported to be three times as much as that of rice. . . .

And it is observed that the most common use of the seeds, all over the world, is in the form of cakes or sweetmeat balls prepared by binding the popped seeds in white or brown sugar syrup.

So what do we truly know? Sauer, on the one hand, tells us that the amaranth originated in the New World; an Indian scientist tells us, with equal emphasis, that Asia is the plant's birthplace. The differences are only compounded by the fact that the two debated locations are on precisely opposite sides of the globe.

Perhaps the answer is more likely to be discovered in the words of botanist Edgar Anderson, who explores the question of plant origins in his catalytic book, *Plants, Man and Life:*

> *How many of our dooryard plants trace back to Asia, which ones joined up with us in Europe, what few of them are authentic Americans? Most of these questions we cannot answer precisely. Yet these questions are only the first part of what we want to know in understanding the relationship of man and plants. In the lands where these fellow travelers joined our entourage, what led them to join up with us in the first place?*
>
> *In many of our western states one drives for hour after hour, and sometimes day after day, between long lines of wild sunflowers which, all untended, border the highways. In Idaho, in Wyoming, in the Dakotas, in Kansas and Nebraska, one frequently sees the double line of yellow leading all the way to the distant horizon. Yet the grasslands on either side of the highway may be virtually free of sunflowers. In a dozen western states the highway winds through a grassy or shrubby landscape between parallel rows of sunflowers. What is there about Man which makes him unconsciously adopt such plants as the sunflower? What is there about sunflowers which permits them to succeed along highways or in railroad yards or on dump heaps, but keeps them away from many native grasslands?*
>
> *It is not until one sits down to work out precise answers to such questions that he realizes that unconsciously as well as deliberately Man carries whole flora about the globe with him; that he now lives surrounded by transported landscapes; that our commonest, everyday plants have been transformed by their long association with us so*

that many roadside and dooryard plants are artifacts. An artifact, by definition, is something produced by Man, something which we would not have if Man had not come into being. That is what many of our weeds and crops really are. Though Man did not wittingly produce all of them, some are as much dependent on him, as much a result of his cultures, as a temple, or a vase, or an automobile. . . .

An ancient, transported flora covers the rolling hills of coastal California. They are draped with a distinctive grassland which differs in makeup and in origin from our eastern meadows. Almost none of it was introduced deliberately; it, too has followed Man around the world, but by a different route. . . .

The bulk of the plants in these coastal grasslands are not originally Californian. Many of them have been there since before the days of the forty-niners, but they trace back to another part of the world with a similar climate and much older civilization. They are Mediterranean weeds and grasses that started moving in with the earliest Spaniards and swept over the landscape, at times almost obliterating the original vegetation. The native grasses still persist here and there; most of the beautiful wild flowers are native, but the bulk of the vegetational mantle is a gift, or a curse, perhaps both gift and curse, from the ancient civilizations around the Mediterranean sea. The plants which are growing unasked and unwanted on the edge of Santa Barbara are the same kind of plants the Greeks walked through when they laid siege to Troy. Many of the weeds which spring up untended in the wastelands where movie sets are stored are the weeds which cover the ruins of Carthage and which Americans camped in and fought in during the North African campaign.

How did these Mediterranean weeds get to California in such numbers? We have little exact information, but it is not hard to make a reasonable guess. As soon as livestock was brought in, the weeds traveled in the hay and in the seeds of field crops. Probably the introduction began with the very earliest Spanish visitors. When the sailing ships were loaded in the Old World, their supplies would have been stacked up on the quay. Every time this was done, a few little pieces of mud could have become caked on kegs and boxes or caught in the cracks. Most weed seeds are small. Hundreds of them could have traveled in every shipload. Of these hundreds, a few lodged in the proper sort of spot when the ship was unloaded. California's climate is very similar to that of Spain, and in those days there would have been few native plants fitted to survive in the strange scars Man makes on the face of the earth. The weeds brought in by the Spaniards already had much experience of Man. Some of them had evolved through an entire series of civilizations, spreading along with Man from the valley of Indus to Mesopotamia and on to Egypt and Greece and Rome. Some had long histories behind them before they ever reached Spain, and for hundreds of years had been selected to fit in with Man's idiosyncrasies. . . .

Fennel, radish, wild oats — all of these plants are Mediterraneans. In their native countries they mostly grow pretty much as they do in California, at the edges of towns, on modern dumps and ancient ruins, around Greek temples and in the barbed-wire enclosures of concentration camps. Where did they come from? They have been with Man too long for any quick answer. They were old when Troy was new. Some of them are certainly Asiatic, some African, many of them are mongrels in the strictest technical sense. Theirs is a long and complicated story, a story just now beginning to be unraveled, but about which we already know enough to state, without fear of successful contradiction, that the history of weeds is the history of Man. . . .

Yet the commonest plants are the least known. . . .

CHAPTER THREE

It is our business to learn how to make an effective record of these plants in Man's transported landscapes, and we will probably be most efficient by not sticking too closely to some hard and fast formula in starting out on this quest. For example, on my last visit to Honduras, we set out to study the native gardens at the homes of small farmers in the upper ends of the valleys. These are little, dooryard plots, mixtures of orchards-vegetable-flower gardens. . . . The first one we studied extended the world's understanding of the grain amaranths. Though these plants had not been previously reported from Honduras, we found this little garden was peppered with the attractive, bright red plumes of the amaranth. Inquiry revealed that the family was not growing it for seed, and did not even know it could be used for nourishment.

To them, it was a magic plant, whose seeds, wrapped up in little packets and carried close to the chest, had the power to ward off a cold, or cure one after it had started. . . . We are planning to make friends with several of these cottagers and make as complete a record as possible of all their dooryard plants, ornamentals, vegetables, fruits, and weeds. It will be an accurate record, not only of the plant itself, but of its relation to the family, why it is there, what they do with it, and how it fits into their lives.

Taken together, these excerpts from Sauer, Singh, and Anderson portray a mysterious, fascinating, and diverse portrait of the amaranth — one which raises almost more questions than it answers. There is such an aura of uncertainty, of doubt, of supposition and indefinite conclusion in these words of three scientists that instead of being irrefutably informed, we are, instead, irresistibly provoked into theorizing on our own.

After all, when one considers the tone of supposition throughout these reports by three men of science, there would seem to be as much integrity in theory as there is in "the facts." Consider the phrases that punctuate the scientific reports like ink blots on an old manuscript: "Unsolved puzzle. . . . faint trail. . . . fragments of information. . . . cryptic statements. . . . no details. . . . often confused. . . . probably. . . . evidently. . . . little exact information. . . ." Can there be any doubt that the amaranth's complete history has yet to be written?

Yet, one theme resounds through the history that has been compiled. It is stronger, louder, more consistent and less varying than any of the others. Like the trade winds, it blows constantly from one direction across the globe, century after century.

Wherever you look for information about the amaranth's times past, you find that this modest plant which evolved from one of the planet's most ubiquitous weeds is endowed with more than merely nutritive assets. The amaranth, we are told by every observer, has a history of feeding the soul as well as the body.

Its flowers are still placed on the altars of cathedrals in Spain. English colonists sailing for America carried amulets of the seed with them; later, Victorian Americans would claim that the plant attracted lightning. Mexicans of ancient times made idols from amaranth seed for their religious ceremonies; after those ancient deities had been replaced, they molded rosary beads from amaranth dough. Hopi Indians in North America have a centuries-old tradition of using amaranth dough as a traditional food for ritual celebrations. Hondurans grow the plant as a magic medicine, Inca cultures used it in their complex religious pageants. The Chinese call it millet from heaven, and in India it is known as the immortal grain.

There are, we are told by taxonomists, five parts of the planet that were the centers of diversity for cultivated plants. The places include Central America and northern South America on the Pacific coast, south-central Europe, central Africa just north of the equator, southern India, and central China. Is it any coincidence that each of these locations turns up whenever researchers begin to trace the amaranth's trail back through the centuries? I don't believe coincidence can explain it; nor do I dismiss the incredibly consistent commonality of the ancient amaranth's metaphysical overtones — a character trait unlike those of any other grain.

Are you aware of any other foodstuff that is known by its metaphysical as well as its physical dimensions around the world, across time and through cultures as diverse as the Hopi Indian and the Szechwan Chinese? Like me, you probably can't think of any, and it is this exceptional aspect of the amaranth that must be one of the primary clues to history; and, as Edgar Anderson explains, the history of man as well.

The supposition that a single plant could be universally identified as metaphysically significant goes beyond the province of mere circumstance. There is nothing in the plant's appearance, as dramatic as that may be in some species, that would identify it transculturally as a plant of mysticism. There is, in my opinion, no way the amaranth's kinship to the spiritual could have made itself known through the plant's natural presence. From the caves of Arizona sandstone to the settlements that cling to the sides of the Himalayas is too long a span for happenstance.

How, then, did the amaranth acquire its global identification as a companion of the spirit as well as the body? Someday, perhaps, the answers to that question will be known. Until then, any of us, using the few building blocks of history so diligently assembled by the likes of Jonathan Sauer, can construct our own scenarios, using logic and likelihood as the mortar that holds those building blocks in the patterns of our own design.

For me, that design begins before the Aztecs and the Incas. There is enough truth in the records of Montezuma's reign to establish the veracity of Tenochtitlán as a capital city of incredible grace and complexity for its time. As more and more evidence is unearthed (of which, more later) it becomes archeologically reasonable — if not demonstrably provable — that there was indeed, as Montezuma said, a civilization that preceded the Aztecs. The Aztec leader of 1520 was not reluctant to portray his cultural predecessors as wiser, greater, and richer in all ways than the civilization he ruled. For a monarch as proud and as untouchable as Montezuma, such admissions would not have been made had not the truth of that parent culture's greatness been unimpeachable.

There are historians and investigators of the Mexican pyramids who present detailed and serious arguments as to the origins of the people who built them. Don Carlos de Sigüenza y Góngora, a professor of mathematics at the University of Mexico a century following the Conquest, was also a poet, astronomer, historian, and geographer. Born in Mexico, the son of a Spanish noble family, Sigüenza cultivated friendships with the native Mexican Indians, persuading them to bring forth manuscripts, paintings, and carvings which they had kept hidden for fear of being burned at the stake as heretics.

It was from these documents and a carved calendar that Sigüenza, with his penchant for mathematics, was able to calculate the year 1325 as the year the Aztecs had built Tenochtitlán. Before the Aztecs, Sigüenza wrote, there were the Toltecs, and before them, the Olmecs who came "from the East." Sigüenza was convinced Atlantis had existed, and he argued that emigrés from that lost continent had been the forebears of the Olmecs. In turn, he believed, Atlantis had been peopled by explorers from the even more ancient land of North Africa where the pyramids of Egypt stand as duplicates of those built ten thousand miles away in Central America.

The Sigüenzas of history have been with us always. His manuscripts were confiscated when he died, in accord with Inquisition custom, and most of them destroyed. We could have learned from them; but it is quite doubtful if we would have learned the answers to the mysteries of the pyramids. Like the huge structures of the Incas, the intricate calendar wheels of the Aztecs, the celestial data recorded in the burial rooms of Egyptian tombs, and the secrets of China still locked in urns of emperors from dynasties no longer remembered, there are vast evolutionary puzzles still waiting to be solved.

And, as long as those puzzles exist, there will be romanticists and scientists, and journalists and botanists, trying to put the pieces of this planet's earliest inhabitants together.

Given some of the ancient artifacts now being unearthed from the very ruins of Tenochtitlán which Cortez ordered buried beyond recovery, there can be solid cases argued for the truth of the hypothesis that pre-Aztec cultures had calculated the circumference of the earth, that they had long known the planet was round, that they had laid out longitudes and latitudes, and accurately plotted the movements of the sun, moon, and stars across the heavens — knowledge which would have given them every logical reason for attempting the globe's circumnavigation. There is even, in some of the more theoretical interpretations of the artifacts, the basis for believing that there was intercontinental communication of a kind, millennia before Cortez thought he had "discovered" a primitive society in the place he called "New Spain."

As I ruminate on the common characteristics assigned the amaranth by the places in this world that were the seats of those ancient cultures — Peru, Mexico, China, India, Africa — I am convinced that there was communication and visitation. For me, the presence of common metaphysical qualities assigned a plant in places tens of thousands of miles apart is a more convincing argument than the similarities of pyramidal construction or the presence of definite Chinese influences on certain pre-Mayan artifacts.

We are talking about what Edgar Anderson calls "diffusion," a process of communication that evolved throughout the globe because people, somehow or other, were able to travel that globe. As Anderson discusses it in his book:

> *To what extent were there cultural contacts across the Pacific in pre-Columbian times? The archeological and ethnobotanical evidence is not clear on this point. One set of anthropologists, the diffusionists, feels that there is strong evidence for the early spread of high cul-*

> *ture from southeastern Asia across the Pacific to the New World. Another set of experts, equally eminent, maintain quite as stoutly that there could at the most have been no more than rare or sporadic contact, and that the cultures of the New World are matters of independent invention. There are able anthropologists both among the diffusionists and the inventionists. European anthropologists on the whole have been more in favor of transoceanic diffusion; American anthropologists are more likely to explain our indigenous high cultures as a flowering of the aboriginal American intellect. A generation ago, when leading American experts were more uniformly antidiffusionist than at present, a witty English anthropologist referred to this theory as the "Monroe Doctrine of American anthropology."*
>
> *A number of anthropologists from both camps have been quick to see that in the argument of diffusion versus invention, the evidence from cultivated plants is critical. You may independently invent a loom, but you can't independently invent an Asiatic cotton. . . .*

Nor, in my view, can you independently bestow qualities of spiritual as well as physical nourishment on the amaranth, right down to the details of how symbolic food from the grain is prepared. The same communication that allowed identical celestial calendars to be developed in pre-Columbian Central America, the Middle East, and South Asia, was the system that allowed the lore of the amaranth to girdle the globe of those ancient cultures, and to survive to this day among the "primitive" peoples who, as it turns out, may not be as primitive as we assume, but instead, are the preservers of a way of life that was much more highly civilized and developed than our own.

In my scenario, using the same building blocks of established facts history and the scientists have given us, and assembling them in the same spaces of hypothesis that the unknowns allow, I see the amaranth assuming its place of importance for very real, down-to-earth reasons. It began as a weed, unlikely to grow naturally anywhere except along river banks and other exposed places. As early Man developed, as the cave dwellers began to establish their communities, with more bare ground, with dumps, and with other likely amaranth habitats, the weed moved itself into human "caveyards."

Its arrival was noticed almost immediately by a species whose hold on survival was tenuous, but stubborn. Man was not likely to ignore a possible food source which moved itself into the very approaches to his living space. So the amaranth leaves were tasted and eaten. Even today, you will find that the young leaves of the most common of the weed amaranths are not only edible, but tasty. Then, perhaps centuries later, the grain was processed and cooked and baked. We don't know precisely when Man discovered the practice of milling and mixing grain; we do know it was one of the first "processed" food forms.

It was after that, I believe, that the wisest of the observers among the early tribes noticed that the amaranth eaters were thriving. The plant's unique combination of nutritional components encourages growth; perhaps it was this basic aspect of keeping people physically stronger that first prompted tribal leaders to ascribe "heavenly" qualities to the amaranth. If it made humans grow larger (i.e., more like gods) then it follows that the grain must be the food of the gods.

I am also of the opinion that somewhere along the lengthy and wavering line of evolving civilization, the amaranth saved a people from calamity. I visualize a flood, an earthquake, a glacial avalanche which destroyed whatever vegetation had been in place. The amaranth, ubiquitous and favoring barren, open ground, took root and spread. The survivors of the disaster, having no alternative, began eating the plant to survive starvation. Not only did they survive, but they grew healthy. Again, in those times, such serendipity could only have been viewed as a miracle, a gift from the gods.

Later, as the early cultures developed, word of the amaranth spread across the populated centers of the globe: the same places where the environment sponsored food production, and thus the nourishment for cultural advance. These times are the spaces in history, the vacuums of knowledge unfilled as yet by what we call facts. All we do know is that there were, indeed, highly developed civilizations that have since been buried by time, leaving behind the looming puzzles of the pyramids, the mysteries of Stonehenge, the question marks of ancient icons, temples, statues, obelisks, glyphs, a few crackling manuscripts written in indecipherable codes, and the great stones taken from temple floors, carved with obscure creatures, designed with an intricate ingenuity that is still beyond our power to interpret.

For these civilizations, wherever they were, the amaranth was a staple grain, a potherb, and a part of the religious ceremonies on which the ancients built their cultural permanence. It could grow almost anywhere, could be tended with the most basic tools, produced large amounts of grain in relation to the cultivation space required, and, in and of itself, the plant supplied much of the nutritional minimums needed for growth, mental alertness, and good health.

It was fate and history that brought Hernán Cortez to the shores of the one place on the earth where a people still maintained traceable, even strong, links with those past cultures.

We could (and hopefully still can) have learned a great deal from the Aztecs. But, in what has since evolved as a typically western attitude, Cortez considered the Indians "primitive" and therefore to be valued only in relation to the gold that could be stolen from them, and the land that could be taken in the name of Spain. Methodically and mercilessly he set about destroying the culture he had found. He slaughtered the Aztecs, burned their crops, dismantled their capital city, and, in the process, cut the links which might have allowed wisdom to flow from time past to time future.

In the autumn of 1978, a group of construction workers in

Mexico City unearthed a ten-ton, circular stone carved with the image of the Aztec moon goddess Coyolxauhqui. The stone had been part of the central temple of Tenochtitlán — destroyed and buried 458 years before by Cortez and his mercenaries. The discovery prompted Mexican archeologists (one of them named Eduardo Moctezuma) to cordon off some 18 thousand square yards of the choicest real estate in downtown Mexico City. In spite of the protests of merchants and real estate developers, the work of finding more artifacts, of unlocking more secrets, of putting back together what Cortez took apart, is continuing. Part of the city, at least, is in the process of being returned to its original builders.

Is it possible that this latest discovery will tell us more about the amaranth's time past? Perhaps. We shall have to be patient; it will be years before the relics are interpreted and understood.

What we have already learned from this and other discoveries of recent years is that the term *primitive* has been sadly misused and abused. Indeed, it was, as it turns out, Cortez who was primitive in his sightless destruction, and not the culture that he and Spain destroyed. A crop grown by some of the most highly developed people of time past is not a crop to be forgotten merely because we live in a later age. Time does not necessarily improve man's intelligence merely by its passage. Instead, time leaves much that we need to know buried in the wreckage of conquest and exploitation.

Right now, in this hungry world, in this time present, we need to know more about the amaranth and some of the other, older crops that "primitive" people cultivated for centuries. And, just as it did millennia ago, information about the amaranth has begun to circle the globe. Time past has become time present, and the amaranth has survived to be part of it. For that, the people of time future are likely to be grateful indeed.

Chapter Four

The world of agriculture in the four centuries since the amaranth was destroyed as a major food source in Central America has undergone startling and complex changes. The most startling and the most complex have occurred in the last hundred years and only now are many agriculturists, scientists, economists, and nutritionists beginning to perceive that not all the changes may have been for the best.

The patterns of contemporary agriculture coincide with the patterns of the Industrial Age; and the doubts now being raised, the new directions now being considered are products of the postindustrial era. We are, in this second half of the twentieth century, in a time of transition. Those at the vortex of change, those whose lives are continually affected by the shift in social and economic values are seldom able to perceive the precise natures of those changes. We are too much a part of the movement from one era to another to be able to pinpoint just where we are going. All we know with certainty is that there is a transition in process and that we are part of it.

There are signposts along the way, however, and to understand the nature of the agricultural transition, we can look first at recent history — a history, by the way, well documented and quite clearly defined. The recent history begins little more than a century ago, about the time of the Civil War. Indeed, it was a weapon that first established a new way of doing things — a way that ushered in the Industrial Age.

That weapon was the repeating rifle, and it was the inventor, Eli Whitney, who designed a system of uniform parts for the gun — parts which could be manufactured separately and then collected and assembled by relatively unskilled laborers in a central plant. That system, and it is difficult to realize now that it was one that had to be "invented," was the start of mass production.

But instead of being remembered for his catalytic creation of a way of manufacturing that was to change this country and the world, Whitney is remembered primarily for the cotton gin, probably because that is a machine we can still use, still look at, and still attribute to a specific time in history. The theory of mass production, on the other hand, is an abstract entity — not something you can build a model of, or a machine you can crank with a handle.

Nevertheless, it is infinitely more important. What Whitney, and later mass producers like Henry Ford, began turned out to be nothing less than a way of life for an entire nation. For in the systems of mass production, there are also the foundations of an American socioeconomic value structure that has grown for more than a century and now has an effect on the way much of the world looks at itself and its problems. Mass production's beginnings are the beginnings of the Industrial Age, and the value profiles shaped by that age have been applied to almost every American endeavor of the past hundred years, with agriculture foremost among them.

Those value system profiles — reduced for the sake of argument to their most basic expressions — can be said to include the following characteristics of the Industrial Age. It is an age which began by exploiting readily available, nonrenewable energy resources which the earth had been banking for millennia. Coal, oil, and natural gas practically exploded from the ground as the mass producers looked for the low-cost energy that would fuel their efforts. At the turn of the century and for three generations thereafter, nearly everyone was convinced that these energy sources would always be cheap and available. They planned their enterprise around that basic assumption.

As technology rocketed from one discovery to the next, powered by the heady knowledge that raw energy was so wonderfully at hand, mass production outdistanced the hopes of its most enthusiastic believers. With the arrival of the automobile on the American scene, the nation's course was set for the next 50 years. We developed, as politicians so often told us, a socio-economic system better than any the world had ever known. More people in this country enjoyed a better standard of living than any people before them in any part of the world. "No American need ever go to bed hungry," was the aphorism that most frequently described the enviable American condition.

In its essentials, that system pivoted on these basics: it was (and in many ways, still is) capital- and energy-intensive; what had been a labor-intensive way of manufacturing became less and less so. In addition to high-energy demands, the Industrial Age also required that mass production be paired with mass consumption and mass waste. Without the latter two, the former would have led only to great mountains of produced goods with no market. Thus the advertising industry was also spawned. Exhilarated by what they perceived as solutions to the problems that had plagued Man through the ages, the leaders of the Industrial Age became convinced they should also do whatever technology could accomplish to keep natural presences out of the provinces of Man. From this came air-conditioned skyscrapers without windows, chemical pesticides, dams and water projects that rerouted entire river systems, and earth-moving enterprises that could, and did, literally move mountains.

In brief, Industrial Age premises rested on these basic foundations: it was (is) an era of mass production, mass consumption, and mass waste that moved toward centralized, high-energy systems that were (are) in essential disharmony with nature. There is no judgment inherent in this analysis; it is merely a statement of the facts of the Industrial Age. Much that is good, as well as some effects that are proving to be not so helpful, has evolved from the past century.

We are, however, in the midst of discovering that some of our complete faith in the system may have to be revised. We are learning at this end of an era that we may have swung to some extremes which can better be remedied by moving toward a balance between what we learned from preindustrial times and what we have discovered during the past century. We are embarked upon a postindustrial journey.

The essential patterns of that journey into the future can be traced through any of the fundamental activities: agriculture, education, economics, government, religion, and human relations. The principles that apply are the same. The story of agriculture becomes the story of banking or presidential elections, medicine or the cosmetics trade. Each moves from a preindustrial phase, to an Industrial Age, and is now in transition to a postindustrial time.

What happened to agriculture, here and around the world, as a result of these changes, however, affects more of the world's people than what was happened in every other area of human endeavor. Freedom from hunger is still the most sought-after freedom on the planet; it is a freedom as yet unattained by half the world's four billion people. Indeed, there are probably more people hungry in the world today than there were 5 years ago, and certainly more than there were 20 years ago.

Long after the death of Cortez, in preindustrial times, the globe's people survived on an agricultural system that was decentralized, labor-intensive, using low-energy, community-based cultivation and methods that were essentially in harmony with nature. As a result, the number of crops used to supplement and support life around the world were as different as the natural surroundings where they were planted. Diets, as they had been for millennia, were widely varied. As many as 50 different sorts of plant crops fed much of the world's people.

As America led the way into the Industrial Age — and, as

much of the globe's breadbasket, set global dietary patterns — agriculture became the opposite of what it had been. Instead of being decentralized, with plots and home gardens in nearly every community and hamlet, entire states were plowed to grow hundreds of thousands of acres of corn, or wheat, or potatoes. Instead of low-energy, labor-intensive, varied agriculture, we entered a time of high-energy, capital-intensive monoculture. The small, community-based farm all but vanished as agriculture became agribusiness, subject to the same economic laws of mass production, mass consumption, and mass waste as any other American industry.

Instead of 40 or 50 crops around the world, as there had been in the mid-nineteenth century, we entered the mid-twentieth with much of the entire world dependent on no more than 12 crops: corn, wheat, barley, soybeans, potatoes, rice, millet, sorghum, oats, rye, peas, and peanuts constitute the bulk of the crops that stand between the planet and starvation. Monocultured with vast applications of chemical fertilizer and pesticides, tilled in square-mile expanses by costly machines that consume large amounts of petroleum products, the crops are subject to waves of pests who breed in direct relation to the monocultural plenty of whatever they prey on.

The battle against pests must always be fought by Industrial Age monoculturists; there is no alternative. When farmers operate in defiance of nature, they are locked in a never-ending battle with natural forces. If, some year, one or two of those battles should be lost, especially in the fields of the more important crops — like corn and wheat — the world would suffer intensely. On the other hand, crop success is almost as unfortunate. When pests and the weather do not conspire against Industrial Age agriculture, the result has often been surpluses of such magnitude that their distribution cannot be managed. Even though 100 thousand of the world's children die from malnutrition and starvation every week, surplus corn and wheat rots in American farm country. Sometimes entire streets of farming towns are blocked off while hills of grain are piled along them, adding to the irony already inherent in the sight of so much food being wasted on the doorsteps of middle Americans who are told almost every day that they must share what they have with the less fortunate peoples of the Third World.

On the surface, such imbalances are puzzling. It is not easy to understand why surplus corn grown in Minnesota is piled in school yards and tennis courts, while the same newspapers that tell us about that also report riots in the marketplaces of Bangladesh because people can no longer bear the pangs of the starvation their entire nation is suffering in the wake of recent floods. The inequities become even more confounding when we are told one American farmer produces food enough for five hungry mouths.

Why can't the expertise that allows the American farmer such efficiency be exported? The answer has to do with the nature of the Industrial Age agriculture. A typical, large midwestern American farmer with 35 hundred acres to work needs these pieces of mechanized equipment: four 15-ton trucks, three pick-up trucks, seven tractors, three center-pivot irrigators, and three wheat combines that cost $30 thousand apiece. That is high-energy, capital-intensive farming. The farm, as large as it is, is run by the farmer, his family, and two hired hands. The crops are basically two: wheat and sugar beets, and the business of marketing is so complex, the farmer must spend hours each day with a computer.

It would not only be foolhardy, but impossible, to try to export that kind of equipment and the expertise required to manage it to a country like Bangladesh. But, in many ways, this has been American policy whenever the world food crisis has been addressed. The nation does make concerted efforts to export tractors, to export the entire Industrial Age approach to agriculture. Or, if tractors do not seem appropriate, the Congress votes foreign aid funds, with the unwritten stipulation that the money be used to build a tractor assembly plant, to be set up and supplied by American manufacturers. Or, the Congress votes to send agronomists to other parts of the world to assist with the further advance of what has come to be known as the Green Revolution, basically an extension of American Industrial Age agricultural philosophy that leads to dependence on one or two crops that must be planted on a large, centralized scale, nourished with fertilizer and protected with pesticides.

Yet, we are told by James P. Grant, president of the nonprofit Overseas Development Council — an organization with a worldwide reputation for skill in understanding and helping with the problems of developing nations — that more people are hungry and malnourished today in the world than were hungry and malnourished five years ago.

It is the realization of such truths, along with some equally shattering truths about the status of the world's energy resources that has begun — ever so slowly — to alter opinions and practices in American agriculture. We are, as a nation, a long way from ending Industrial Age farming; on the other hand, the process of building a foundation for the postindustrial agriculture that is certain to evolve has already begun.

Oil has been the primary catalyst for precipitating the transition. The recognition, just a few years ago, that petroleum is indeed a nonrenewable, finite resource, led many Americans to consider a concept they had never before even visualized as a possibility. They began to ponder a concept of limits.

In the minds of many, although they may not have realized it, the seeds for such thinking were planted when the first rockets left Cape Canaveral for outer space, and later for landings on the moon. The photographs of earth that came back to us from those incredibly sophisticated technological achievements were photographs that moved us from the Industrial Age — at the height of its technology — toward a postindustrial future. For the first time in the history of Man, we were presented with graphic, irrefutable evidence of the smallness and fragility of our planet. We could cover the photograph with a silver dollar, yet we also covered ourselves, our nation, and both seas that stretched from its shores. For many people the planet became, for the first time, "Spaceship Earth." Looking at its roundness, at the recognizable outlines of continents, the pale blue of the oceans, and the delicate embroidery of the clouds, many of us felt the beginnings of a kind of anxiety at the smallness, the aloneness of what had become, in comparison to the rest of space, a tiny place for humanity to call its only home. The limits of the planet were there for all to see.

They came to mind several years later when we had to wait in line at service stations, wondering all the while if there would be enough gas when our turn came to fill the tank. We didn't understand at first; we thought the entire episode had been foisted on us by an industry looking for ways to raise the price of what had been such a cheap and reliable source of energy. Many people still buy that reasoning, but whether they do or not is relatively unimportant alongside the deeper realization that all share: the Industrial Age America of Henry Ford will never be the same again. We are, for the first time since the early settlers — who had a precise understanding of nature's laws — beginning to understand that there is no such thing as a free lunch.

We sense that something has gone awry with our energy supplies; either there is a real shortage, or an imposed one. Whichever it is, we get less oil and natural gas and have to pay more for it; our cars are getting smaller and our mortgages larger. The cost of an average American home is nearly $60 thousand and climbing. Natural resources, once wasted with such abandon, have acquired a new sort of value. A nation's coal, oil, steel, copper, coffee, and cotton become the keys to instant wealth, great power, the ability to buy armies and air forces. There is, more and more of us recognize, only so much to go around, and whoever has what little is left is lucky; the rest are out of luck.

This view of the world is brand new; it imposes a concept of limits on a nation which has operated since the nineteenth century as if there were no limits. Now some of us are anxious about a future we never before contemplated. What happens, for example, to that midwestern farmer's three combines and three tractors when he can't get all the oil and gas he needs to run them; or — and this is far from a hypothetical question — what happens to the price of food when the cost of producing that food takes the inevitable jump it will take when a farm run by machinery gets charged a dollar a gallon or more for the fuel it takes to operate all those machines. If a farm is run on a high-energy, capital-intensive system, then it is bound to be severely affected by sudden increases in the cost of equipment and energy.

What the concept of limits faces us with are questions like this: If the cost of oil continues to climb, can we afford to eat carrots in Maine that are grown with machines in California and transported by machines across the country? No matter what the advantages of centralized, chemicalized, large-scale farming, they can be quickly obviated by an increase in the costs of the energy it takes to distribute the crop from one central location to the rest of the nation.

American agriculture, it now appears, as we enter the postindustrial age, will have to be decentralized as resources become more and more limited, more and more valuable. As it turns out, this is not a prospect for alarm. Instead, just the opposite is true. As more and more nutritionists are learning, and more and more experts concerned with feeding the world are beginning to believe, the sooner America begins to balance its agricultural equations, the sooner all of us (the people of the world included) are likely to be eating more, and better. Instead of continuing Industrial Age agriculture, we will move into postindustrial farming — a time when the best of the preindustrial and the industrial eras are combined, when the wisest technologies are blended with the oldest skills.

To understand how a postindustrial farm can work, you have only to correct the excesses of an Industrial Age system. If a midwestern farm in the 1970s operates with centralized, high-energy, capital-intensive methods that are in fundamental disharmony with nature, then its postindustrial counterparts (and there will be numerous smaller farms replacing the single large one) will be decentralized, varied, low-energy, labor-intensive places that are managed in essential harmonies with nature.

As this occurs, I believe all of us will be pleasantly surprised at what happens to our health, as well as what happens to food prices — which will be generally more stable, less the victims of inflation. And, as for nutrition: there are already those experts who will argue convincingly that as we get further from mass-produced, monocultural agriculture that pivots on the ready availability of pesticides and herbicides we will move closer to longer, healthier lives; smarter, healthier children; and lower medical costs.

There are many examples in time past of how the small, decentralized, community-based farms can work. I think of a place named Zion Hill in Maine when I try to visualize the agricultural communities of the future. Zion Hill is difficult to find these days; a first-time visitor — as I was on the day I discovered it — has to be guided to the top of the ridge that bears the name. The thriving settlement that was there at the turn of the century has been abandoned so long that thick woods have covered almost every trace of human habitation.

But there are clues to the past still rooted in the soil. Fruit trees — bent, gnarled, and black with their years — grow in rows among the disorderly profusion of the conifers and birch that have taken over fields that were once plowed every fall and spring. If an explorer looks closely in those woods, he can find remnants of roadways, log bridges, and stone markers. Cellar holes where homesteads once stood yawn in the forest floor, filling slowly with wild raspberry and alder. Ancient lilacs reach for the sun over the tops of pasture pine, marking dooryards that are no longer there.

There was a thriving settlement here, a self-sufficient community of committed people who learned how to work with the land so it gave them sustenance and purpose in exchange for their labors. None of today's farmers would ever have picked the location as a cost-effective place for agriculture. The ridge is steep; just beyond it are mountains, their slopes covered still with some of the nation's wildest forest. This is not a place for easy living. The soil is thin, spread in a stingy layer over the glacier-gouged granite just inches beneath the forest duff. Markets are distant, reachable over lonely, twisting roads.

The people who came to Zion Hill were driven; they came in search of solitude, of a privacy formidable enough so they could practice their particular beliefs without interference or public notice. But they had to eat, and what few neighbors there were, down below the ridge where the land begins to flatten, had no food or creatures to share.

So the citizens of Zion Hill became farmers in a place where farming might never have been attempted by anyone who thought he knew more about its risks and realities. Perhaps it was the enthusiasm and zeal the Zion Hill people applied to their farming, perhaps it was merely because they had never been told that theirs was a bad farming choice, but whatever the reasons, they not only produced food for themselves, but consistently harvested more than they needed.

The milk and dairy products from their cows, their cheeses and butter, their eggs and poultry, the wool from their sheep, the vegetables from their gardens, and the fruit picked from the rows of apple and pear trees still marching through the pines like old soldiers — each of these and more, the Zion Hill people loaded into wagons and took to markets in Augusta and Skowhegan, a day's wagon ride away. They sold the produce for good prices; so much care had been taken with its husbandry that the mill workers of those mill towns would buy Zion Hill foodstuffs first — they looked better and tasted better. So not only did the Hill people learn to treat the grudging ground with the dedication needed to make it productive, but they did their jobs well enough so that the same hard ground a professional farmer of today would scoff at returned them extra benefits — benefits they could sell so they attained economic as well as nutritional independence.

CHAPTER FOUR

Zion Hill is now nothing but a memory, a place where autumn winds sigh in the pines and cellar holes open — mute mouths waiting to speak the history of this all but forgotten settlement. Whatever the fabric of the community, it was torn by time. Younger generations abandoned the beliefs that had held the place together, they came down from the ridge to the towns of Harmony and Wellington in the valley, and they let the forest grow back where it had once held undisputed sway.

Although they are built in less remote and foreboding places, thousands of Maine's small farms — and many of the nation's — are also Zion Hills. Trees grow in the formerly fertile fields, barns that once sailed across the meadows like great, gray schooners are wrecked on the reefs of their own granite foundations, and farmhouses fall into the neglected indignities of their own cellar holes. Yet, like the Hill, each of these places was once a producer — a place where crops were raised without the dust of pesticides, the mist of herbicides, or tons of packaged chemical fertilizer.

It is important to remember that these places can each — every one of them — be producers again. Granted, it is difficult, as many dedicated people of all ages and backgrounds have learned in the past several years, to breathe new life into these farms that were buried so long ago. It takes a singular and determined effort to rouse an abandoned farm from a sleep as deep as death; but it can, and has, been done.

In my view, as the Industrial Age becomes a postindustrial one, it will need to be done, and with some urgency. I foresee the time in the not too distant future when the knowledge needed to efficiently, carefully, and lovingly restore deserted land will be a skill much in demand. Books will be written on the subject; periodicals will be published and will be quite successful. The reason: there will be an informational vacuum waiting to be filled.

As that midwestern farmer with 35 hundred acres and a small army of mechanized, high-energy equipment begins to compute his taxes, his fuel costs, the costs of acquiring new machinery, and the replacement of worn-out machinery with human labor, he will find that he must charge such high prices for his produce that there will simply be no market for it. Just as they have already begun taking charge of their own homes and making them less dependent, American consumers will look to the low-energy, small community farm to help them meet their food needs, with no sacrifice in quality, and at a lower price.

The Zion Hills of America will be restored; the spruce, pine, hemlock, birch, alder, oak, and maple that have invaded the fields that once were, will be cleared and cut to cordwood, their stumps will be uprooted in an echo of the stump-hauling done by colonial settlers with their oxen. Varied crops that will never be sent across the country, but instead will go only to the neighborhood, will be planted and harvested, more with hands than horsepower. Nothing of what we have learned over the past century will be wasted, but much of what was once thought to be a chapter from our past will be revived and relived in the future.

When that happens, crops will change, as they have already begun to change. Corn that grows well in Indiana and is designed to be harvested by huge machines is not likely to also grow well in Maine where it will probably be harvested by hand. The crops will fit the needs and realities of their growing environment, as well as conforming to labor-intensive, postindustrial harvesting techniques. Instead of processing grain through the stomachs of beef cattle to produce a costly and inefficient sort of protein, much more vegetable protein will be eaten directly as it comes from the soil. And as beef and the other mass-produced crops of the seventies begin to lose their clout in the marketplace, public tastes will shift to a more varied diet, a more vegetarian diet, and a diet that looks to the Zion Hills of our past for guidance as well as excitement.

The great majority of today's supermarkets are, after all, more masters of display than they are maestros of variety. All those shelves so creatively lined with their brightly (and expensively) packaged goods are illusions of variety, nothing more. A careful tour of ingredients, rather than brand names, will keep turning up the same basics: corn, wheat, sugar, potatoes, rice, citrus fruits, bananas; produce like celery and lettuce from the vast truck farms of California; pork, beef, and chicken in the meat cases; various dairy products; breads, beer, and soft drinks. Take these items, or basic ingredients, out of any typical American supermarket, and you'll find you won't have much that's edible left inside.

This is modern, convenience shopping — an Industrial Age benefit which we have been persuaded is an integral part of the "good life" in America. As it turns out, that term is proving to be an ironic misnomer. Increasing evidence is being assembled which points to the typical American diet of animal protein, fats, starches, sweets, and alcohol as a diet likely to shorten life rather than make it better. For although we are said to be the best-fed nation on earth, the designation would be more accurate if it specified that we eat a great deal, but are not as well nourished as some people who eat less. We tend to be obese or overweight, short of breath, susceptible to heart and liver ailments, and — although the final evidence has yet to be verified — more likely to get cancer than any citizen of the Third World, primarily because of cancer-aggravating chemical residues in our processed, mass-produced foods, and in the air we breathe.

As I said: there is no such thing as a free lunch; and, as someone else has said, we are what we eat. The mass-production Industrial Age farm has made certain kinds of food popular and readily available. These easy-come, easy-go foods, however, are not proving to be as good for us as they might. We are the most developed, richest, most powerful nation in the world — fueled by nonrenewable resources and a socioeconomic system shaped by Eli Whitney and Henry Ford — but we are also a nation of pudgies who die, on the average, as early, or earlier than many people in other parts of the world. You might well wonder what "we hath gained" in the course of living the Industrial Age to its fullest.

The resulting discussions will soon prove academic, however. As the Industrial Age wanes and the postindustrial era begins in earnest, the high-energy, centralized, monocrop, large farms that service national and international markets will be replaced by low-energy, decentralized, varicrop small farms that service community and regional markets. The plants that are cultivated will lend themselves to labor-intensive harvesting and processing, will grow well with a minimum of care, will have generally high yields, and will be planted in relation to how well fitted they are to a particular environment, rather than how well they can fill the demand which advertising has created for a particular popular taste.

Before you begin to become anxious about the possibility of being denied your T-bone; your white, sliced bread; your sugared cereal; and your hot, buttered corn-on-the-cob, you might be more optimistic and meditate on the wide variety of new taste sensations that will come your way as a result of the agricultural forces for change already irresistibly at work.

What will happen, among other changes, is that many of the older . . . some say "forgotten" . . . crops will be restored. Others suggest that our diets will become significantly different, more varied, more replete with perfectly good, and some better, nutritional foods that are now ignored and often wasted by a food-glutted society. Some signs that this process has already begun can be found in the fishing industry.

Depleted by hordes of large boats of all nations, the fishing grounds off New Zealand — long recognized as the most productive in the world — have failed to produce significant crops of haddock, cod, and herring in recent years. These, the most popular food fish, have suffered the full force of the technological, Industrial Age, fishing fleet whose depth recorders, fish finders, and other electronic gear enabled it to find the fish, so huge, high-powered gear could be employed to catch them. "Fished out" is the commercial man's term for what has happened on the Grand Banks to the herring, haddock, and cod. Although the entry of foreign vessels has been restricted in some ways by the establishment of an American two hundred-mile territorial limit, the United States Department of Marine Fisheries has had to impose strict catch quotas on United States fishermen, and even with those quotas, fishery biologists are wondering if the threatened species can ever be restored.

As a result, many kinds of fish that were formerly tossed back into the sea are being utilized. Dog fish, whiting, cusk, hake, and other species — some without popular, "marketable" names as yet — are no longer tossed to the gulls and crabs, but processed and marketed to a public that has become accustomed to buying whatever is put on the shelves in a handsome, convenient package.

In this case, the public is gaining more than just a better-educated palate. It is gaining a more varied diet, and every study made of nutrition and its effects on health and longevity indicate that a varied diet is a large improvement over one that pivots on just a few ingredients — no matter how "good" those few ingredients may be for you, in and of themselves.

For an example of just how varied the diet of a "primitive" culture (one without supermarkets) can be, consider this "shopping list" compiled by research scientists* who studied in detail the diet habits of a twentieth century tribe of healthy, long-lived Australian aborigines:

twenty-nine kinds of roots
four kinds of fruit
two species of cycad nuts
two other types of nuts
seeds of several species of leguminous plants
two kinds of mesembryanthemums
four sorts of gum
two kinds of manna
flowers of several species of *banksia*
six sorts of kangaroos

*Modified from J. B. Birdsell, in the *American Naturalist* 87:171-207, 1953

five marsupials somewhat smaller than rabbits
nine species of marsupial rats and mice
birds of every kind including emus and wild turkeys
three types of turtles
eleven types of frogs
young of every species of bird and lizard
twenty-nine kinds of fish
four kinds of freshwater shellfish
four kinds of grubs
two species of opossum
dingoes
one type of whale
two species of seals
seven types of iguanas and lizards
eight types of snakes

Which is not to imply that you rush out to your backyard and start harvesting snakes which you then convert to a stew that will prevent any possibility of heart attack. The point is, as scientists and nutrition researchers are increasingly learning, that there are some scientific advantages to so-called "primitive" diets which strongly indicate that in terms of illness prevention and general good health, old and varied diets which utilize plant and animal life native to the region are better for you than the contemporary American diet — generally unvaried, and the product of one part of the nation made available to all the rest.

And just precisely what is that American diet doing to Americans? One nutritionist — Barbara Ford — answers that question beyond much argument in her book, *Future Food: Alternate Protein for the Year 2000:*

> *In recent decades, a number of scientific studies have uncovered a sinister link between the American diet and the American way of death. Or, to be more exact, between the high consumption of fat, particularly animal*

fat, and sugar on the part of populations of North America and Western Europe and the major killer diseases in those areas. The fat- and sugar-rich affluent diet, as some scientists refer to it, is a goal to strive for in the eyes of most of the world but it may be the principal reason for early death in developed countries. . . .

In our nation, as in most of Western Europe, the big three among killer diseases are coronary heart disease, cancer, and stroke. Coronary heart disease leads the list in our country. A general term for diseases involving the coronary arteries through which the heart supplies itself with blood, coronary heart disease often involves a "heart attack" in which the supply of blood to the heart is cut off. Although the incidence of coronary heart disease is dropping in the United States, it still takes some seven hundred thousand lives every year. Cancer is the second leading cause of death, stroke the third in the United States. Stroke results from an impaired supply of blood to parts of the brain and also involves the arteries. In both coronary heart disease and stroke, the mechanism that cuts off the supply of blood to the heart or brain is the deposition of fat on the inner wall of the arteries, a condition called atherosclerosis or, more popularly, hardening of the arteries. Included in the fatty deposits is a fatlike substance called cholesterol which is naturally present in the blood and other organs but which is also acquired from foods of animal origin.

Recent studies indicate that in young American males, the deposition of cholesterol in the arteries begins as early as the age of two. By their twenties, what one United States researcher calls "massive infiltration" of the arteries by cholesterol is underway. . . .

Most of the research attention on the link between heart disease and the presence of fat in the blood has been focused on cholesterol but another type of fat is now believed to be associated with heart disease: triglycerides.

Triglyceride, which is simply the scientific name for fat, is the predominant fat in both our diet and our own body tissues. Research now indicates that elevated blood levels of both *cholesterol and triglycerides are linked with heart disease.*

What does all this have to do with diet? . . .

Some of the most provocative studies involving cholesterol have been carried out with populations that have moved from an area where a low-cholesterol, low-fat diet is popular to an area where a high-cholesterol, high-fat diet is in vogue. In one study of this nature, Japanese who moved from Japan, where fat consumption is about one-third of that in the United States, to either Hawaii or California had a higher incidence of coronary heart disease than the Japanese population in Japan. The rate of increase was moderate in Hawaii but high in California, where the diet of the new immigrants underwent the biggest change. Further evidence of the effect of the affluent diet is the fact that coronary heart disease incidence in Japan, although still low, has tripled over the past 15 years, a period that corresponds to a rise in meat eating.

The links between dietary fat and sugar, on the one hand, and blood fat levels and atherosclerosis, on the other, have been known for decades, but research carried out within the past few years now implicates dietary fat in some common forms of cancer in this country, too. Dr. Ernest L. Wynder, president of the American Health Foundation, a private research organization with offices in New York City and Valhalla, New York, believes our daily diet probably plays a much larger role in cancer than food additives, substances which are often condemned as cancer-causing agents.

The new evidence on which Wynder and a number of other scientists base this belief associates a high level of dietary fat with an increased incidence of colon (large bowel) cancer and breast cancer. Colon cancer is the sec-

ond leading cause of cancer death among United States men, breast cancer the leading cause of cancer death among United States women. Studies show that in countries where the population eats a diet with a high fat content, the incidence of both colon and breast cancer soars. Conversely, colon and breast cancer incidence drop in areas where comparatively little fat is eaten. The graphs for geographic incidence of the two diseases are almost identical, with the same Western nations and nations with Western culture, such as New Zealand, clustered near the top, the same Asian and South American nations grouped at the bottom.

At or near the top are Canada, the United States, New Zealand, Australia, Denmark, and the Netherlands, all among the leaders in meat and dairy-food consumption. At the bottom are Thailand, Japan, the Philippines, Colombia, and Ceylon, all with diets based primarily on vegetable protein and/or fish. Most of the nations with a low incidence of colon and breast cancer are developing countries with a low level of industrialization, but Japan is heavily industrialized. Japan's inclusion in the group of countries with low incidence of breast and colon cancer indicates that some factor other than those associated with industrialization must be responsible for the difference in death rates from breast and colon cancer. Genetic predisposition? No, because studies show the incidence of colon and breast cancer rises in Japanese who emigrate to the United States and adopt a Western diet. Also, groups such as the Seventh-Day Adventists in the United States who practice vegetarianism have less breast and colon cancer than the rest of the United States population.

An environmental factor must be at work, and the only one researchers have been able to single out is diet. Americans eat three times as much fat as the Japanese and our fat intake, unlike theirs, is mainly of animal origin. . . .

Although a casual relationship between the high level of fat or protein in our affluent diet and the leading killer diseases isn't established as yet, many nutritionists and other scientists involved with nutrition believe that diet presents a clear "risk factor." Their advice: Change the affluent diet now. *If we wait, they argue, it will be too late to save many American lives. . . .*

What changes do these experts recommend?

Less meat, fewer dairy products, less fat of all kinds, less sugar, less salt (a risk factor in heart disease). More vegetable protein sources, more low-fat animal protein sources such as veal, chicken, and fish, more starch carbohydrates, substitution of some unsaturated fats for saturated fats. Specific recommendations include most of the alternative sources of protein or their by-products, particularly grass-fed beef (less fatty than grain-fed beef), low-cholesterol meat and cheese analogs, vegetable protein-carbohydrate sources such as grains, beans, seeds, and roots, and highly polyunsaturated oils, for instance, safflower and sunflower oils. One group of foods that emerges with new luster is the starchy carbohydrates, those old no-no's of a thousand slimming diets. Now it appears the starchy carbohydrates have been unjustly maligned. New evidence indicates that starch not only does not make you fat but that a diet rich in starchy grains, seeds, roots, and beans (many of which, of course, are also rich in protein) can also make you healthier than a diet rich in animal fats.

It is, however, one of the world's most telling ironies that while many Americans literally "eat themselves to death," millions of people across the oceans are not able to find enough to eat to properly sustain their health and their lives. Consider what Barbara Ford has to say on the subject of malnutrition:

The food shortage is principally a problem of what are known, today, as developing countries (a polite phrase that has replaced the former undeveloped countries). Most of them are in Asia and Africa. At frequent intervals, some of these nations experience severe shortages of both protein and calorie sources, and death rates mount. Even in good times, malnutrition is a chronic condition among a large segment of the population in developing countries. This past year, 1977, was one of those good periods. The world experienced its second year of bumper harvests and food stocks were high for the first time since the early 1970s. India, one of the poorest developing countries, amassed a 20-million-ton reserve of grain, the chief source of both protein and calories in that nation. Few people died of hunger in 1977. Between 1971 and 1975, however, the death rate climbed by two million, an increase attributed to starvation or diseases associated with severe malnutrition by the Worldwatch Institute, a private nonprofit research institution with offices in Washington, D.C. Most of those who died were children. And for every person who died, scores or even hundreds suffered various degrees of malnutrition.

Another research group, the National Research Council, which is affiliated with the National Academy of Sciences, estimates the current number of malnourished people in the world today, during a period of bumper harvests, at between 450 million and 1 billion.

Nutritionally speaking, malnutrition is simply a state in which the body does not receive what it needs from the food ingested. One of the causes of malnutrition is too little food but it can also be caused by an imbalance of one or more kinds of nutrients such as protein. Severe malnutrition — what physicians refer to as clinical malnutrition — encompasses a number of familiar and not-so-familiar diseases including pellagra, scurvy, and rickets. Since children require about twice as much protein and calories as adults, malnutrition strikes them more often. The most severe form of malnutrition, protein-calorie malnutrition, is a childhood syndrome. It is manifested in two diseases, kwashiorkor, *which is caused by a protein deficiency, and* marasmus, *which is caused by a calorie deficiency. If untreated, both can and do lead to death. According to one estimate, 10 to 20 million young children in developing countries have severe forms of* kwashiorkor *or* marasmus *at any one time.*

For most of the world's hungry, though, malnutrition takes the less dramatic form referred to by physicians as subclinical malnutrition. Victims show symptoms such as apathy, weakness, low weight, short stature, inability to handle stress, overexcitability, and lowered resistance to disease. Again, children are affected more severely and frequently than adults. Subclinical malnutrition isn't directly lethal, but it sharply raises the odds of early death from other causes. In one survey carried out in Central and South America, 57 percent of all child deaths were found to be associated with nutritional deficiencies. The symptoms of kwashiorkor *and* marasmus *are unmistakable, but parents of malnourished children often fail to realize that their youngsters' apathy or weakness is due to diet. "Neither the children nor their parents realize they are sick because they do not know what it is to be well," says Dr. Sohan Manocha, a biologist at Emory University in Atlanta, and author of* Nutrition and Our Overpopulated Planet *(Charles C. Thomas, 1975). Based on surveys in various underdeveloped countries, Worldwatch Institute estimates that subclinical malnutrition affects as many as one-half to two-thirds of the children in underdeveloped countries. . . .*

In recent years, the presence of large numbers of malnourished children in developing countries and smaller numbers in developed countries has taken on a new and frightening significance. Intelligence in children, new studies indicate, is linked with malnutrition, particularly protein malnutrition, not only in the children themselves but in their mothers as well. . . .

CHAPTER FOUR

Protein is of such vital importance in early life because it is the substance required by the body to build cells, as well as to maintain and repair them. Contrary to popular opinion, babies and pregnant women need much more protein than athletes. A severe illness or operation also calls for extra protein. The basic unit of protein is the amino acid, 22 of which are known. Of these, 11 cannot be made by the body; they can be obtained only from food. These proteins are known as the essential amino acids. They must be given to the body in certain amounts. A protein source that supplies all the essential amino acids in the correct amounts is called a complete protein. If you eat a modest amount of it, it supplies all the protein you need. . . .

But will there be enough plant protein to go around in the future to provide the optimum mixture of amino acids? Will there be enough of even a fairly good source of plant protein like wheat or rice?

The world population explosion indicates that the answer to both questions may be no. Consider a few figures plucked from various reports. The world now has about 3.92 billion people. The population is growing at the rate of about 64 million people per year. By the year 2000, the world population is expected to be about 6 billion. Asia and Latin America, between them, will have about 80 percent of the world's population by then; India alone will have about 1 billion people. Some of the more pessimistic prophecies project 50 billion people on earth less than a century from now. The United Nations, however, forecasts a stabilization at between 10 and 12.5 billion by 2025 to 2035. If this more conservative estimate is borne out, there will be two and one-half to three times the population that now exists on earth 50 years from now!

Food supplies, many agricultural experts fear, cannot keep up with the population. "We are attempting to deal with finite limitations of resources in the face of what appears to be infinite human potential for procreation," said J. George Harrar, president emeritus of the Rockefeller Foundation, in 1974.

We have learned, thanks to the clear prose and the thorough research of Barbara Ford, that here in America we are doing ourselves little or no good (and quite likely are doing harm) with a protein-rich diet that is also overloaded with fats and sugar. Meanwhile, two-thirds of the world badly needs the proper sort of protein if its children are to grow up with the intelligence to manage developing nations, and if malnutrition is to somehow be stopped from taking its deadly toll of tens of millions of children and adults around the world.

The development of new foods, then, that can help keep Americans in better health here at home, while, at the same time, they help solve the problems of malnutrition in the world's developing nations, would logically appear to be a matter worthy of high-priority attention from agricultural leaders of both the private and public sector. Happily, that attention is currently being paid, and with increasing and encouraging intensity.

The battle against world hunger in the less-developed nations, and the effort to improve the nutritional intake of overfed millions in the industrialized countries, is a simultaneous effort. What is learned on behalf of Asia and Africa, it turns out, is also proving to be beneficial to West Germans and Americans. Increasing numbers of botanists and agronomists are combing the poorest, semiarid regions of the earth, looking for wild strains of wheat, sorghum, millet, pigeon peas, yams, cassavas, and other "forgotten crops."

Among the many institutions coordinating the work around the globe are: ICRISAT, an acronym for International Crops Research Institute for the Semi-Arid Tropics, established in 1972 in India; IRRI, the International Rice Research Institute in the Philippines; the International Maize and

Wheat Improvement Center in Mexico. Two other institutions, located in Nigeria and Colombia, concentrate on tropical crops. There is also a potato research center in Peru, an animal health laboratory on Kenya, a livestock center for Africa in Ethiopia, and "dry area" centers in Lebanon and Syria. The Consultive Group on International Agricultural Research (CGIAR) is the umbrella organization that coordinates the work of ICRISAT and the other groups.

Their basic effort was expressed in a few words by the director of ICRISAT, Leslie D. Swindale of New Zealand: We are working on behalf of the farmer whose crop is his life. If that crop fails, the farmer and his family starve.

"Our job," says Dr. Swindale, an agronomist, "is to help the poorest segment of humanity. We must concentrate on subsistence crops, those that are least known and studied, the ones the Green Revolution never touched."

And while ICRISAT works in the fields of Asia and Africa, scientists in America are assisting with basic research. In 1978, the annual report of the National Academy of Sciences — produced by the National Research Council, the Academy's working arm — includes a summary of the work done by a Council committee of agricultural scientists who have spent the past several years researching "new crops." The summary report on that work, written by Roger Revelle, one of the nation's leading agricultural scientists, points out that only 30 of the hundreds of thousands of plants on the earth now supply more than 95 percent of the calories and protein consumed around the world.

The scientists on the committee identified and catalogued more than four hundred plants not now in general use that could be converted to cultivation in tropical and semitropical countries. Of these, 36 "new crops" were selected as being especially promising in terms of edibility and nutritional value. Some of them were attractive because they are "leguminous" — they make their own fertilizer by extracting nitrogen from the air. Other crops were found to yield an abundant harvest in a short growing period. Still others get a high rating because the food they produce can be easily and inexpensively stored or preserved for extended periods.

An editorial in the *Saturday Review* (25 November 1978) suggests that the National Academy study is a "historic document, worthy of the kind of attention given the Club of Rome's 'Limits of Growth' report published in 1972. For if the academy's report is translated into active programs by tropical nations of Asia and Africa, the world may yet be spared some of the horrifying hunger disasters that have been so widely predicted."

There is a pattern, you see, to the agricultural and nutritional research and testing being done around the world. And although many of the individuals involved (including the *Saturday Review* editorial writers) may not yet have made the connection between what can be done to help the Third World and what needs to be done to improve the health of most Americans, that pattern will also evolve in the near future.

It will because there are people like Barbara Ford documenting the dangers of the "sweet and fat" American diet, while, at the same time, agronomists like Swindale and Revelle discover and report on the nutritional value of crops that could help ease hunger pains around the world. Taken together with the concept of limits now being recognized by more and more Americans, the trend presages a major shift in agricultural values. Instead of petro energy monoculture the hallmark of Industrial Age farming — the crops of the future will not only be more nutritious, but will be grown on a different value scale altogether.

To repeat a set of standards for future crops described earlier in this chapter: as the postindustrial era begins in earnest, the high-energy, centralized, monocrop, large farms that now service national and international markets will be replaced by low-energy, decentralized, varicrop small farms that will service community and regional markets. The plants that are cultivated will lend themselves to labor-intensive harvesting and processing, will grow well with a minimum of care, will have generally high yields, and will be planted because they fit a particular environment, rather than how well they fit the needs advertising has created.

There is an irony here — an irony which began with Hernán Cortez more than four hundred years ago. Cortez looked at the new worlds and ancient civilizations he discovered with typically Western attitudes. He did not see the Aztecs as a people of subtle and complex culture, but rather as primitives to be conquered and enslaved. Nor did he see the land we now know as Mexico as a place to be husbanded, to be cared for, to be eased into an era of wise resource use. Instead, he saw it as a place to be plundered. His awards for heroism were earned with the accumulated treasures of a nation that had been developing for tens of centuries. Cortez not only stole those treasures, but he saw to it that the land was ravaged for whatever resources could be easily ripped from it.

Cortez is no exceptional explorer; he is, instead, a metaphor for every Western adventurer . . . the men who won the west . . . navigators of uncharted waters . . . wagon train leaders, trappers, scouts, the Davy Crocketts, the Wild Bill Hickoks were no different. For them, as for all the adventurous men of the days of exploration, the purpose was to locate resources, to take territories, to subjugate natives. And, once the land had been claimed, exportable, high-value crops were planted. Such were the basics of the colonial value system, the system that evolved in the times when there were not only no recognized limits, but when every resource was viewed as immediately exploitable. Indeed, it would have been heresy to suggest otherwise. Every native population was primitive; every crop of theirs was of little value. It was only the people of the West who understood how to make the world work on their behalf; it was the Western mind that conceived the accumulation of private riches, that established corporations, that created trading companies, that took treasures (and called them prizes), and mined, and plowed, and hewed, and dammed, and constructed until they had built the foundations for the Industrial Age — the most incredible century of material gain and human accomplishment the world had ever seen.

In the exploitation process, the fields of Mexico, the valleys of Brazil, the rain forests of Asia, the plains of Africa . . . wherever Western colonialists set the flags . . . were stripped of the natural crops that had evolved over time and replanted with the produce that could produce wealth: cotton, rubber, tea, coffee, sugar, and spices took the place of the crops native populations had depended on for subsistence. Over the decades, peoples that had been sustaining their numbers for millennia found themselves at starvation's brink. The plots that had once produced household essentials were now planted with coffee, or cayenne, or rubber trees.

What had been "one world" evolved painfully into the First World, the Second, Third, and Fourth, with most of the people living in the Third and Fourth. Because their resources had been diverted, exhausted, or exploited, they became wards of the First and Second; their "starving millions" became the Westerner's burden.

Meanwhile, consuming the fruits of the colonial ethic, millions of Americans, British, Germans, and citizens of the other First and Second World nations became progressively less healthy. The combination of industrial pollutants and the sweet and fat diets that industrial wealth made possible also debilitated the body, and, quite likely, the mind of the same Western Man who had created and carried the Cortez system throughout the untouched places of the globe.

Only recently have there been the tenuous beginnings of a new idea. Some people are beginning to understand, for example, that Americans who live their lives a bit differently, who exercise more and utilize nutritional, less harmful, diets will, more than likely, enjoy active and fulfilling lives that cover a span considerably longer than the current norm. Instead of retiring to nursing homes and hospitals shortly after they retire from their jobs, more older Americans — like homesteading authors Scott and Helen Nearing — would be more likely to plan a new house, and then build it themselves. As a result, one of the nation's more critical problems — what to do about the aging — would be considerably eased, not by governmental formula, but through something as basic as a change in diet and a change in attitude.

That is what the postindustrial ethic signifies — a future that is different, but also potentially better. The shift from a societal value concept based on incremental growth to one that pivots on an understanding of the concept of limits is a transition that need not raise anxieties. Indeed, it should ease them.

However, most politicians and institutional leaders are wary of addressing the issues of the future; we are offered a plethora of proposals for extending the past into the present, and very few for using the best of the past to create a better future. Given the state of the Third and Fourth Worlds and their vast nutritional needs, along with the sorry physical condition of all too many overfed Americans, it seems obvious that there is need for a new approach to the basic issue of food and diet, starvation, malnutrition, and overeating.

What happens to this basic life-support system could set patterns for the postindustrial reformation of every other basic system: education, economics, government, and all the rest. It is interesting to see how those patterns are beginning to fall into place. Having destroyed what there was of the ancient cultures in the continents he discovered, explored, and exploited, Western Man, after five hundred years has looked at the results of his policies in the Third and Fourth Worlds and, after countless studies like the one recently completed by the National Academy, has decided that one of the steps that will help the most is the restoration of what the planners have labeled "forgotten crops."

Thus have we come full circle from the moment Hernan Cortez landed at Veracruz and began the five hundred-year cycle which has brought us to the beginning of the end of the Industrial Age. It was the soldiers of Cortez who destroyed what was then not a forgotten crop, but the nutritional mainstay of a civilization thousands of years in the making. So well destroyed was the plant and the culture that nothing less than a rediscovery in the late twentieth century can restore it, and, at the same time, help end the malnutrition of millions of Third and Fourth World people while it also helps ease the diet excesses of their conquerors.

This is the plant which the National Academy researchers ranked high on their "desirable" list when it came to selecting the species of "new" crops that could help turn back the threat of famine. The plant produces ". . . three varieties of grain containing 15 percent more protein and 53 percent more starch than corn grown on the same amount of acreage." The report of the Academy scientists recommends that it be fully developed in the tropical highlands of the world because the concentrated food value of the grain makes it especially suitable for people in areas where roads and vehicles are scarce.

It also grows well on small plots with a minimum of cultivation, can be harvested with low-energy, labor-intensive methods, and can be raised as a companion to other crops.

This "new" crop, this "forgotten" crop, is, of course, the amaranth, and to state that simple fact is to underscore one of the great ironies of the past five hundred years. We have reached a point in time where the most august American institutions — the National Academy of Science itself — and a host of equally impressive international organizations agree that the amaranth has potential, not only as a crop that can help end malnutrition around the world, but one which can also help correct the misnutrition that shortens the lives of too many Americans. It makes one wonder just how much of true significance has, indeed, been accomplished since Western Man launched the Age of Discovery five centuries back.

Among the few individuals in the nation who might have shortened that time by a decade is John Robson, M.D., a nutritionist who at the time was at the University of Michigan. As a specialist who had researched the plants of many cultures in a search for nutritionally valuable foods, Dr. Robson tried for years to get government bureaus and commercial food manufacturers interested in the production possibilities of the "forgotten crop." Then, in 1972, Dr. Robson wrote a letter which began a series of events that may, in another few years, begin the final reversal of the process Cortez began more than four centuries ago.

Chapter Five

The man Dr. Robson wrote to was Robert Rodale, head of Rodale Press in Emmaus, Pennsylvania — a publishing firm founded by his father. As the publisher of *Organic Gardening, Prevention,* and the *New Farm* — three magazines concerned with crops and good nutrition — as well as considerable annual lists of books on such topics as home gardening, better health through better foods, and other postindustrial topics, Rodale Press was a logical place for Dr. Robson to turn in his continuing effort to arouse more general interest in the amaranth. Neither Dr. Robson nor Bob Rodale, however, could have known on that day years ago where that first contact would lead them and the plant.

"John Robson told me," recalls Bob Rodale, "that he had tried many times without much success to get officials of the large food companies interested in the amaranth's potential. He quite openly admitted he was writing us almost in desperation because all other sources of research help he had approached so persistently over the years had turned a deaf ear.

"Well, that's the news that sparked my interest. I figured that if those responsible for developing the nutritionally inferior foods which are taking over the marketplace felt indifferent to something like amaranth, then amaranth was very likely a healthful, valuable food. At least, it amused me to think so. Naturally, I told Dr. Robson that we were interested in supporting more work with this ancient and remarkable plant."

Shortly after that, Bob Rodale was mailed a brief paper

CHAPTER FIVE

Robert Rodale, intrigued by the potential of the plant, propelled Rodale Press to the forefront of amaranth research and development.

by Dr. Robson. Its topic, as you might expect, is the amaranth and its relation to the contemporary food habits of most Americans. Several years later, the Robson paper is still one of the outstanding works in its field:

> *Although their nutritional value is unquestioned, amaranth and many other wild plant foods make virtually no contribution to the diet of Western Man. What is even more important is that these highly nutritious plants are being decreasingly used in the Third World. In the face of increasing shortages of food in the world, this neglect seems extraordinary and the reasons for the lack of use of these foods deserves attention. So far as amaranth is concerned, it is known that at one time it was an important food but Dr. Sauer has described how the Spanish conquistadores discouraged its growth in their colonies in the Americas and the Old World. This is not the only reason for the abandonment of amaranth as a food; its use, and disuse in human diets, is part of the fascinating history of food.*
>
> *When man first emerged on the earth a million or more years ago, he met his nutritional requirements by gathering his food. At first, most of the food material came from plants, but it is likely that insects and other easily caught animals, such as lizards and crabs, also contributed to the diet. Later, as hunting skills developed, fish and mammals began to play a more important role in the diet.*
>
> *The evidence for these statements has been provided by archeological research but clearly archeology can only be a crude tool for assessing the diet and the nutrition of prehistoric peoples. Although many of the foods eaten, such as leaves, fruits, and berries, could not survive the effects of time, some nuts and seeds were more durable and they have been found where prehistoric populations formerly lived. These seeds, discovered after thousands of*

years, have largely reverted to carbon but nevertheless, they can be identified botanically. It is assumed that they were eaten because they appeared around campfires, in stores, in caves, and also in the refuse pits. Seeds have also been found in an undigested or partly digested state in the dried up feces (known as coprolites) found in caves inhabited by man thousands of years ago. Again it is reasoned that the seeds themselves, or the fruits and berries bearing these seeds, must have contributed to the diet.

Plant remains found by Dr. Kent Flannery of the University of Michigan in caves in the Valley of Oaxaca in Mexico included the nuts, seeds, bulbs, leaves, pods, rind, and stems of 27 different plants. Acorns, onions, cucumber, and a number of Central American plants such as agave and mesquite, opuntia and nance were some of the foods believed to have been eaten. Animal remains were also found and these included deer, raccoon, hawk, owl, dove, quail, turtle, and lizard. In general, there is a lack of information on the role of amaranth in the diet of man in early prehistory, but by 5000 B.C. the archeological record clearly shows that amaranth was being cultivated for use as a food.

The ability to domesticate plants and animals was the start of the agricultural evolution. This phenomenon has continued to this day, but at first it was confined to the selection of a few wild plants and they were then grown around the home. The plants eventually were found to require care involving cultivation of the ground, weeding, and transplanting. As a result of this, the farmers became more and more preoccupied at the expense of time which otherwise would have been spent in hunting or gathering. Thus the manner in which food was being obtained was drastically changed; obviously these changes would have an effect on the diet and ultimately on health.

The influence of dietary change is very difficult to

determine from the archeological record, but fortunately diet and health can be measured in groups of humans who still live like prehistoric man lived thousands of years ago. Such populations include the Australian aboriginal, the bushmen (or San) of the Kalahari desert, and the Tasaday in the Philippines. With the exception of the Tasaday, who are protected by government decree from exposure to the outside world, the effects of social, economic, and political development continue to alter the lives of these simple-living people. On first exposure to agriculture, they follow the transitions of early man and they learn to domesticate plants and animals as we did thousands of years ago. Eventually they adopt the typical mode of life of the peasant farmers who populate the Third World.

Even peasant farmers are affected by the progress of civilization and in recent years the farmers of Africa, Asia, and the Americas are finding themselves exposed to the influence of the Industrial Revolution. This started in Europe and North America in the early part of the nineteenth century and at that time, the discovery of steam, the internal combustion engine and farm implements, and other inventions revolutionized the farming system. Intensive farming of single crops such as corn, or wheat, or rice, became the most profitable way of producing food. This was especially important for consumers who were working for cash in towns, and who were dependent on the rural areas for their food.

It was not anticipated that these advances in technology would have an effect on peasant farmers in the remote parts of the world for many years. This assumption has been found to be incorrect, however. For example, in 1933 in New Guinea, it was noted that the native gardener was found to be "squatting on the ground and digging desperately with a short stick." It was not expected that he would "leap up and land in the seat of a tractor to go thundering over his land with a 12-furrow plough behind him." Professor Oomen has noted, however, that as an improvisor, the New Guinean did leap up and "in 1960 he could be seen sitting on a thundering tractor in the Kumbe region, now and then using his bows and arrows to shoot a kangaroo." Thus the inhabitants of the Third World who formerly lived as simple farmers have now found themselves applying new agricultural techniques involving the use of fertilizers, herbicides, pesticides, and even plant genetics. In their specialization they are becoming increasingly dependent on food grown by others. They are now purchasing foods that have been refrigerated, frozen, dried, cured, pickled, and packaged in a number of different ways. These developments in food production and technology obviously affect the availability of foods and hence the choice of foods eaten.

In these circumstances it is not surprising that growing indigenous plants like amaranth has been neglected and their use as foods is being forgotten. Many will argue that the changes are for the better, that we no longer need amaranth. It is important, however, to appreciate the impact of these changes on man as an individual, or on man as part of a society in the modern world.

As a first step toward understanding the impact of the Agricultural Revolution, it is necessary to study the characteristics and implications of the diet of aboriginal man. Studies of the Tasaday, the Australian aborigines, and others living as we lived ten thousand years ago, confirm the evidence from the archeological record, namely that man consumed a tremendous variety of foods. The foods of the Australian aboriginal included 29 kinds of roots and 11 kinds of frogs. The Dayaks of Sarawak consume 40 different plant foods, the Tasaday, 50. This pattern of food behavior even occurs in this country, for example, Indians of the Sonora desert of the United States utilize 76 species of desert seed plants, including amaranth. At the other end of the climatic spectrum, Eskimos utilize 27 different plant foods, including fermented lichens obtained from the stomach of the caribou.

The effects of the adoption of farming are very clearly shown in the example of the Tiruray in the Philippines. When civilization reached these hill people of the Cotabato Cordillera, some of the tribal groups retreated further into the hills where they continued to maintain their traditional mode of life; others became peasant farmers. The transition from an aboriginal mode of life of hunting and gathering, within the same culture, provides an insight into the impact of social and economic development. The aboriginal Tiruray consumed 104 different kinds of food materials, obtained by hunting and gathering. After adopting a more settled way of farming, the numbers of food materials was reduced to 59.

Simplification in the diet so far as numbers of foods exploited is characteristic of development. The process continues as our modes of life become more sophisticated. The peasant farmer in Africa traditionally grows a number of cereals. In the central part of Tanzania, it is quite common to find sorghum, finger millet, bulrush millet, and maize (corn) growing in the farms. As new methods of agriculture are introduced, the cultivation of the different kinds of cereals is discouraged; as a result, multicropping agricultural patterns are replaced by single crop agriculture, usually based on maize. The process of simplification continues as peasant life becomes affected by the industrial revolution. Within our lifetimes, the diet has become less varied, thirty years ago we ate considerable amounts of wheat and rice, oatmeal, rye, and barley. Sorghum at that time was already considered as fit only for feeding animals. Wheat and rice are now our predominant staples while oats, barley, and rye are consumed only rarely. It is easy to speculate that these lesser-used cereals will be abandoned or relegated for animal feeds in the not too distant future. . . .

CHAPTER FIVE

The significance of variety in the diet can now be appreciated. Nutritionists can plan mixtures of foods so that the amino acid deficiencies of any one food are made up by surpluses of the same amino acids in other foods. With a knowledge of the composition of foods this task is relatively easy but prehistoric man had no concept of nutrition principles, and the same is still true for the aboriginal mother. It is fortunate, therefore, that increases in the numbers of foods consumed increases the quality of the dietary protein. Clearly the greater the number of foods the greater is the chance of consuming a mixture of proteins whose amino acid mix might approximate the ideal protein.

In infancy, variety is absolutely vital for the achievement of an adequate intake of protein. Human beings are mammals and their nutrition in early life is provided by mother's milk. It is assumed that human milk is ideal for human babies and this assumption is borne out by observations on aboriginal and peasant farming communities around the world. In these situations, breastfeeding is very successful and the growth of infants in the first six months of life is comparable to (and sometimes exceeds), the growth of babies in Western societies. Whatever the geographical location, and regardless of racial background, breast milk is inadequate, however, to maintain proper growth after six months of life. This calls for a transition in the diet from one that is exclusively milk to one comprising more solid foods and which characterizes the diet of the older children and adults. This is the weaning period and it is the weakest link in the life cycle of man when many problems face the baby.

Nutritionally, the infant has a great need for calories and protein. In early life, this energy and protein intake is provided entirely by the mother's milk. This is a food offered in close proximity to the mother, and its taste, smell, and consistency became a major part of life. Weaning foods, however, are different in their taste, smell, and consistency, and they have nutritional disadvantages also. For example, the appetite of the young infant is controlled by his calorie intake; as a result the baby stops eating when sufficient calories are consumed. This means that the protein must be highly concentrated, otherwise the baby meets his calorie requirements before the protein requirement is satisfied. The protein content of milk is concentrated in relation to the calorie content but many other foods do not have a high protein-calorie concentration. Moreover they are too bulky for the infant's small stomach.

The aboriginal mother weans her baby on honey, turtle eggs, soft fruits, small pieces of fish, and vegetables. This weaning diet should be compared with the foods given to the infants of peasant farmers in Africa, Asia, or Central and South America. Weaning in such circumstances is based on cereal gruels. If the Western world has made only a slight impact on development, the gruels are usually made from millets, sorghum, maize, or rice. But with increasing contact with the West, maize replaces the millets and sorghum; eventually gruels made from cassava (manioc) replaces the maize. In all geographical areas, the gruel is supplemented by side dishes usually made from leaves. Rarely are beans and peas fed because they are believed to cause stomach upsets. Because of expense, milk, meat, and fish appear very infrequently in a weaning except in communities of pastoralists, or fishermen. The nutritional implications of the gruels and side dishes are important. First, the gruels are bulky; often they fill the child's stomach before he has eaten sufficient calories. Furthermore, the protein concentration is low so that there is little chance of him receiving sufficient protein.

The only hope appears to be training the stomach to accept more food, as a result the mother often resorts to force-feeding. By these means, sufficient food may be con-

sumed to provide enough calories but the protein is still usually inadequate. The question of protein quality also is unresolved. The diet is not varied and it is usually monotonous. With the exception of milk-, meat-, or fish-eating populations, the mother can easily fail to provide the weaning infant with a wide variety of protein sources. This can only mean that the dietary protein quality is inadequate. In these circumstances, it can be predicted that the weaning infant will suffer from protein or calorie malnutrition. . . .

At the local level, attempts have been made to educate mothers in infant care including the preparation of more nutritious weaning foods. Much can be done in this direction, but the task would be very much more facilitated if the emphasis was taken away from Western concepts of what is "good food" and redirected to the best and most appropriate local foods. In theory, amaranth, chenopodium, and a number of other plant foods could all be used. The tasks are first, to find out what foods are available locally, and second, to determine in what form they can be prepared in a suitable fashion for human babies in the weaning period. Certainly from a theoretical point of view amaranth has great potentialities as a food that could contribute to the weaning process. The problem arises when the theory has to be applied in practice. A nutritionally sound food must be acceptable to the infant and above all, it must contain good quality protein in sufficient amounts and concentration to meet the nutritional needs of the infant. Unfortunately achieving the latter is not too easy and at all costs the inherent dangers of using infant foods based on plant vegetable protein as a substitute for milk must be avoided. . . .

While such substitutes for milks are not feasible and may even be lethal, amaranth and other cereal grains have a great potential as foods to be included elsewhere in infant-feeding regimes. For weaning, foods are quite satis-

As a nutritionist, Dr. John Robson saw amaranth's potential as a food crop early in his research.

factory if they contain a protein calorie concentration of 10 to 12 percent. In their usual dry state, amaranth seeds have a protein calorie concentration of about 14 percent.

If the grain could be made into a stiff, thick gruel or dough, the dilution effect of the added water would not be too great but the question of acceptance by the infant would have to be tested. When given as a more dilute gruel the protein quality of amaranth would be superior to maize and sorghum, but nevertheless, the decrease in the protein calorie concentration would greatly affect its value as a weaning food.

In the Western world where nutritional problems are not so great, amaranth could certainly serve as a substitute for wheat, rice, or any other cereal grain. Its protein quality is superior and the main challenge would be finding ways of presenting the grain in an acceptable fashion. As a pablum, it would serve a useful role in the weaning process. Later in life, it could be used as a breakfast cereal or as a candy or dessert. Again it has to be noted that any of these dishes which use added starch, sugar, honey, or any other source of calories would dilute the protein calorie ratio, and this would be undesirable. In adult life, the problem of diluting the protein is not so important and amaranth could take its rightful place in the diet serving as a substitute for other cereal grains.

In summary, therefore, amaranth has a potential value as a weaning food, but its practical importance needs to be explored. Amaranth and other plant seed grains could provide much needed variety to the diet thereby enhancing the quality of the total dietary protein intake. This would be very important in developing countries but research is needed to evaluate different methods of preparing amaranth and the effects of the different methods on the nutritive value of the dish need to be determined. Where storage, marketing, processing, and distribution facilities exist, amaranth could be combined with other indigenous protein sources as a protein supplement to be distributed or sold to mothers of weaning children. Such foods would be comparable to other protein mixtures like Incaparina in Central America, Laubina in the Middle East, Pro-Nutro in Africa, and other preparations made from vegetable protein seeds such as soybeans, chick-peas, and sesame.

In the more affluent parts of the world where the process of weaning does not present so many nutritional hazards, amaranth and other indigenous seeds could be used as a substitute for conventional cereal grains. The problems in this case relate to the supply of the seed and agricultural problems such as harvesting, and difficulties in processing and distributing the food to those most in need of it.

Although several obstacles inhibit plans to introduce amaranth into the diet, these are largely offset by the unique properties of this plant. First, it can adapt to a wide variety of environmental, topographical, and climatic conditions. Its ability to thrive under adverse conditions, and its propensity for rapid and extensive growth under ideal conditions, has resulted in it being classified as a weed. When growing in proximity to conventional crops it could be considered a threat to productivity. Because of this, it is discouraged. In the Western world where intensive agriculture of single crops is practiced, discouragement and elimination of amaranth and other indigenous food plants similarly designated as weeds are customary. It is assumed that they have no contributions to make to food production. It must be remembered, however, that no studies have been undertaken to compare the yield of nutrients (as opposed to weight of crop) per acre produced by cultivation of indigenous plants with the yields from wheat or corn. Certainly from the point of view of energy economics, the ability of amaranths and other indigenous plants to produce high protein seed grains having a high energy content without the aid of fertilizers, mechanical cultivation, and herbicides has tremendous advantages in times of fuel shortages.

In developing countries currently being exposed to Western agriculture the story is different. Rightly or wrongly (and possibly the latter) Western agricultural principles are being imposed on peasant farmers in the developing regions of the world. An evaluation of the impact of these policies on food resources and diets has not been scientifically measured, but the experiences of the Green Revolution have raised many questions regarding the wisdom of applying new agricultural technologies developed in temperate climates to the rigorous climates of the tropics. Such questions are beyond the scope of this discussion, but it must be remembered that amaranths and other indigenous plants thrive well with the minimum of cultivation.

In accordance with the prehistorical facts, in contemporary populations where the environment has not been affected by farming, wild plants continue to be productive and they are able to supply protein and energy for human diets. Once the plants become domesticated, further nutritional benefits accrue. For example, amaranth and similar plants may be grown in clumps and this type of intercropping is very efficient. A single plant of sorghum or maize may be grown together with beans, amaranth, tomatoes, wild berries, etc. The sorghum or maize stalk serves as a pole for the beans to climb and the bean is able to fix nitrogen in the ground. The beans and corn and plants like amaranth collectively provide shade against the hot tropical sun. This allows the more delicate green, leafy plants to thrive and produce fruit. Thus a clump of inconsequential and untidy plant life may be producing in a very efficient manner a variety of nutrients without extraneous energy inputs like tractors for cultivation, herbicides, fertilizers, or insecticides. Such modes of agriculture do not fit into the patterns and principles of modern agriculture, and intercropping systems are actively discouraged. As a result, amaranths and oth-

er plants that not only provided essential variety to the diet but also serve as a famine reserve in times of adverse climatic conditions are kept out of the system.

Over the years, the cultivated varieties of amaranth that may have adapted very well to local conditions fall into disuse and eventually they are lost as a source of food. This lack of knowledge of the past role of amaranths and a lack of understanding of their potential are largely responsible for the current failure of food producers to meet the challenges of increased productivity in developing areas, especially those susceptible to climatic vagaries. It may be postulated with a great deal of confidence that it may be possible to develop different agricultural systems utilizing amaranths and other plant foods, and these may provide at least a partial answer to the recurring problems of the Sahel and other arid areas of the world. Acceptability of these plants is known in many locations, their ability to thrive can be tested quickly. It is also possible that other food plants will be rediscovered, so the prospects for indigenous plants making valuable contributions to world food resources in the future are good. The first need is to replace the present negative attitudes towards these plants with a positive attitude that will enable objective testing of their use to be undertaken.

Difficulties will undoubtedly arise, the problems of harvesting amaranth seeds are constantly voiced and it is implied that harvesting is labor-intensive. This latter argument can be countered, however, by recognizing that developing areas are notoriously rich in cheap labor. Furthermore, it must be recognized that the bird-seed industry successfully harvests other very small seeds. With increasing development and in areas where intensive culture of indigenous plants may be considered a viable alternative there is now evidence that amaranth can be grown successfully. It is possible that other plants will be equally productive. Although crossbreeding occurs be-

tween selected seed amaranths and wild varieties, this problem may not be too serious since at the worst imperfect seeds may still be used for animal feeds. Plant geneticists hopefully will solve this problem so the grain can fulfill its proper role as a cereal and as a nutritive supplement for peoples of all ages.

There are, if you look for them, many questions raised about the amaranth by Dr. Robson; granted, his paper is full of enthusiasm, yet, at the same time, it is a plea for more research, more knowledge, more testing.

What form is best for amaranth preparation after it is grown and harvested . . . what is its real potential as a food . . . what is its practical importance in a corporate world that seems satisfied with the products it is already marketing . . . how many different methods of amaranth preparation have been evaluated . . . what is the reliability of the seed supply . . . what is the actual yield of nutrients per acre cultivated in comparison with other crops . . . is intercropping practical . . . how does the amaranth grow in marginal soil and weather environments . . . what are the problems of harvesting . . . crossbreeding???

Bob Rodale was aware of these questions, and more, when he completed his initial discussions with John Robson, and, as he puts it, "We at Rodale Press quickly plunged into our exploration of the amaranth's possibilities."

By the close of 1974, Joel Elias — an associate of Dr. Robson's — had been to Mexico to search for villages where the plant was still grown. He returned from the trip with seeds, and in 1974 the first grain amaranth plants took root in east-central Pennsylvania on a farm that had enjoyed more than two centuries of productivity, but had never before had the amaranth included in its crop list.

The farm had first been cleared before the Revolutionary War by a family of German immigrants to the New World who settled in the lovely valley. The generations of Siegfrieds

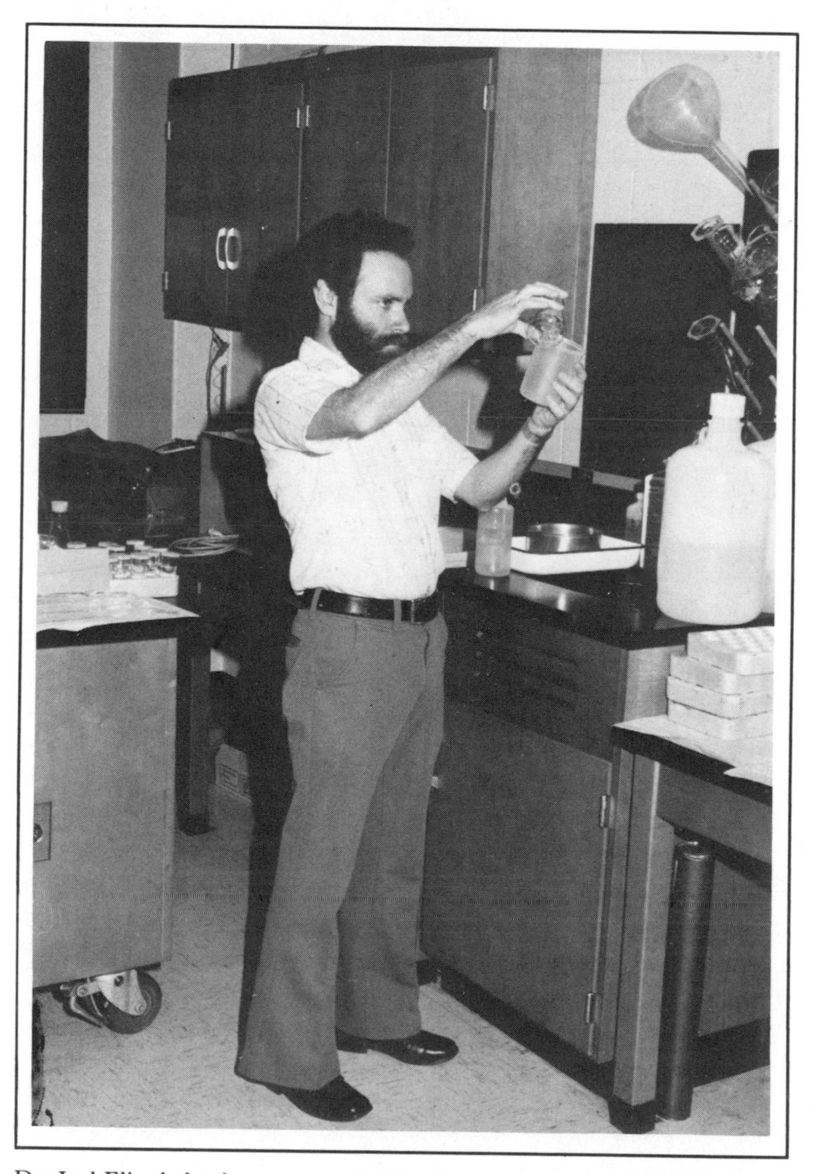

Dr. Joel Elias helped move amaranth from the past to the present, for it was the seeds he brought back from Mexico in 1974 that were first planted at the Organic Gardening and Farming Research Center.

never left the land, and, as the twentieth century began, they had 305 acres under cultivation, had built a large farmhouse and a typically Pennsylvania stone-sided barn as solid and imposing as a cathedral.

Shortly after the farm was purchased by Robert and Ardath Rodale and Rodale Press early in the seventies, it was named the Organic Gardening and Farming Research Center, and some 20 acres of the rich land was turned over to scientists on the Rodale staff as well as some from nearby universities. While a neighbor farmer, Ben Brubaker, tilled the rest of the acreage in the organic traditions of his Mennonite forebears, the 20 acres of the former Siegfried farm were dotted with carefully marked plots, walked over by the shoes and work boots of scores of scientists and agronomists, planted with a variety of experimental and "old standby" crops, covered with

Hay bales shelter a stand of *A. hypochondriacus* from strong winds.

plastic film tents, peered at through microscopes, tested for soil conditions, observed for insect visitations, and generally probed and primped and processed more than any other 20 acres of Pennsylvania farmland anywhere in the state.

Chief among the plants studied at the Research Center for the past four years has been, of course, the amaranth. Not only were the answers to Dr. Robson's questions being sought, but a hundred more problems were posed, a hundred more solutions tested. What began with a relative handful of seeds brought back by Joel Elias in 1974 became, by the summer of 1978, the largest single collection of amaranth germ plasm in the world. Any of the thousands of visitors to the Research Center (and visitors are welcome to join the daily, guided tours in season) could see the rows of amaranth varieties — some with red leaves, some with purple, some with yellow, each waving like bright banners in the summer sun. Other amaranth varieties — among the more than three hundred being grown — bent with the weight of the grain clusters near their tops, and more poked new shoots in orderly rows from the plots where they had already been harvested once as greens for the table. Still more amaranth grew in the shadow of corn rows where it was being watched to see how it works as an intercrop, and through every field walked researchers who bent to study amaranth pollen, amaranth leaves, amaranth color, amaranth resistance to wind and rain, amaranth hybrids, and amaranth species never before planted in this country . . . each of them here in one more effort to find the answers to questions first raised by Dr. Robson, and to the scores of others put by the scientists themselves as they become more familiar with this plant so few of them had met before they came to the Research Center.

Other efforts to learn more about the amaranth are underway at Berkeley, California, where the Cereals Research Unit of the United States Department of Agriculture has begun its own investigation of the grain species, and also at the University of California at Davis, Philadelphia College of

A. hypochondriacus being harvested at the Research Center in 1977.

Pharmacy and Science, New Crops Research Division of the USDA in Beltsville, Maryland, and at Cornell University, Oregon State, Iowa State, Pennsylvania State, and Michigan

State. Numerous graduate students have made the plant the subject of their degree work, and at least four American seed companies include amaranth seed in their catalogs.

There is, however, no more intense effort being made anywhere than the one mounted by Bob Rodale and his publishing/research complex. More than 14 thousand readers of the Rodale magazine — *Organic Gardening* — volunteered to grow amaranth in their backyard gardens in all of the 50 states and Canada. Like the researchers on the old Pennsylvania farm, those volunteers were trying to learn more about the plant, to define its variables, to discover its pollination habits, the dimensions of spacing for optimal growth, and harvest and seed-cleaning techniques. In addition to keeping track of these botanical details, the home gardeners also tested the amaranth's tastiness, both as edible table greens made from the leaves of young plants and a variety of cereals, breads, and other foods made from amaranth grains.

Reports from those volunteers (of which, more later) are on file in the farmhouse room that serves as the nerve center of the Organic Gardening and Farming Research Center. The man behind the modest desk in that office that looks out over the fields has the ruddy complexion and rugged frame of a working farmer. He walks through the Pennsylvania fields with a farmer's long stride, and he talks about the land and crops with the knowledge and affection every true farmer feels for these essentials of his trade.

Richard Harwood, Ph.D., likes being called a farmer because that's primarily how he sees himself. He is more, however, considerably more. As an agronomist whose profession has taken him to Asia and other nations of the world where he helped lead efforts to conquer hunger with innovative agricultural techniques, Richard Harwood's reputation as one of the leading international experts in his field preceded him when he agreed to become the director of the work being carried on by the scientific team at the Research Center.

During the several years that he has been there, much of

CHAPTER FIVE

Dick Harwood's time has been given to the amaranth. It was the opportunity to work with the plant, in fact, that helped bring Dr. Harwood to the old Siegfried farm.

> *The chance to work on amaranth as a breeder played no small role in my decision to join the Rodale staff. I compare the plant to a diamond in the rough. We have glimpsed its brilliance in the great potential it represents, and we want to exploit that potential to its fullest.*
>
> *The work that's accomplished in the Americas with amaranth will not only have importance here, but to many Third World countries with more limited resources, such as the protein-poor, cassava-growing nations of Africa where the crop could make a major impact.*
>
> *In order for science to have the most immediate and effective application (so contemporary thinking goes) it should be developed within the historical, social, and physical environment to which it is expected to apply. The setting provided by the Research Center is one of a growing operation — a working farm, if you will — having a goal of excellence in scientific research and in the organic technique. I am confident that there are few better places to help answer the many questions that need to be answered before the amaranth can become a significant global crop once again.*

Among the desirable characteristics that will help the amaranth take its place as a "global crop" are: uniform germination, flowering and maturing; similar types and positions of seed heads; strong root systems; insect and disease resistance. Amaranth plants to be selected for seed stock should produce a great quantity of golden grain, high in quality protein, and a loose enough seed head to prevent molding. Amaranth selected for greens should have palatable leaves, high vitamin and mineral content, and a large overall yield.

To reach these goals, Dick Harwood directs an effort fo-

Above is an unimproved bulk population of grain amaranth from Mexico. Note the varied heights and irregularity of these plants. Contrast this with the photo below of an improved stand made from a selection of the bulk population. This stand is more uniform in size, is shorter, resists seed lodging, and resulted in 80 percent higher yields than the unimproved stand.

cused on increasing and inventorying the Research Center's collection of amaranth seeds, observing growth characteristics of new lines and promising types from other trials, learning about amaranth pollination habits, and — most important — learning the mechanics of amaranth reproduction.

It is the center's long-term goal to reach the point where an amaranth can be bred that will combine all the good traits Dick Harwood and his coscientists have observed over their years of study, while, at the same time, they reduce the traits they consider undesirable. Once that is done, seed can be produced that will provide gardeners and farmers everywhere with attractive, manageable, and nutritional crops — better, perhaps, than those that once covered so many thousands of acres of preconquest Mexico.

And how much longer will it take for that kind of seed to "germinate" in the research farms and laboratories of the scientists, agronomists, botanists, and researchers who are working around the world to meet the challenge? The best answers to that question can be given by the people involved in the effort, and I began contacting them early in 1979. My first stop was Dick Harwood's office — the one with the window that looks out over the fields of amaranth growing at the Organic Gardening and Farming Research Center farm.

Chapter Six

"It is," said Dick Harwood on a cold, windy January morning, "much further than you think from there to here."

As he said "there," the director of the Research Center pointed out the window to gently rolling Pennsylvania fields where the most intensive effort in the nation had been made over the past five years to test different varieties of grain amaranth. As he said "here," Dick Harwood gestured toward the warm farmhouse kitchen where a pot of coffee perked softly on the stove.

With that one sweep of his arm from field to kitchen, he had framed a process so complex, so linked to so many different aspects of contemporary American culture that there must be few people in the nation who can accurately measure just how far it is from "there to here."

The social, political, agricultural, and economic dynamics of popularizing a "new crop" (which is what the amaranth is in America, even though it has been part of the history of past millennia) combine to create a pattern of change which touches the lives of each of us. In an age of centralization, of agribusiness, of a Washington-based economy; in a time of growing concern for nutrition, of increasing government regulation of the farm and food industries; and in this era of specialization and monoculture which has led to a nutritional dependence on less than a score of the hundreds of crops that once were farmed, the process of getting amaranth grain from those rolling experimental fields in Maxatawny and into the mouths of millions of Americans — and more millions around the world — is a process of startling dimensions.

Among the few organizations which has taken a comprehensive look at it is Soil and Land Use Technology Inc. (SALUT), a consulting firm in Columbia, Maryland. Their 1978 report, funded by the National Science Foundation (NSF), takes a close and detailed look at 54 plant species proposed by more than one hundred selected for more study, and, of that handful, 5 were nominated as the subjects of a long-term, full-scale experimental effort. The grain amaranth was chosen as one of the final 5.

Although the SALUT report to the NFS is primarily an examination of the reasons which sustain the amaranth's selection, it is also a clear blueprint of the course that must be taken to move a crop — any crop — from the field to the dining table. As the report puts it: "The feasibility of introducing a new crop depends not only on solutions to technical agronomic and economic problems, but on the availability of a functioning production-marketing-consumption system." That's a short sentence, but it reaches from "there to here."

Using the report as a kind of blueprint for action, here are some of the steps that must be taken within the established systems for making the journey.

First, the reasons for beginning must be established. In the case of new crops, those reasons pivot these four basic assumptions: (1) the genetic pool of the world's 20 basic agricultural crops may be vulnerable to disease and blight unless it is diversified; (2) most of these present crops are demanding on their environment or can only be adapted to a narrow ecological niche; (3) the crops now depended upon need high energy inputs in the form of fuel, fertilizer, pesticides, processing, and irrigation; and (4) the lack of alternative crops in the marketplace leaves farmers vulnerable to price instabilities over which they have no control.

Most of us can agree with those assumptions; the facts to

support them are available and rather widely known and acknowledged. There are, then, these and more good reasons to take an official look at the amaranth.

What happens next? The first step is understanding how the principals within the agricultural establishment — government agencies, farmers, and marketers — view the introduction process. In their eyes, the adoption of a new crop is a long and gradual development. It starts with the introduction of an exotic species or the identification of an indigenous one, and the growth of a few specimens under trial or experimental conditions. It takes years to complete the agronomic research, to develop varieties, to grow the crop under practical farm conditions, to test the product and its processing methods, and, finally, to develop a market.

And, like every "new kid" in the neighborhood, the new crop has to make room for itself, has to find a place to grow. Sometimes this isn't easy, especially if the "old boys" on the block are reluctant to give up any of their turf. Each of the major crops now grown in the country is raised on the best land available for each of them. If the new crop must have some of that same land, it is immediately in competition, not only with the old, established crop, but with the rising pressures for other uses of the desirable land: housing, highways, and industrial development. Thus, the most successful candidate for a new crop would be one which can use land that is beyond the adaptation range of the established "old boy" crops; the new kid moves in, but he makes himself at home in a place that hadn't been used before. Instead of displacing an established crop, he enlarges the neighborhood.

Looking at the potential for a new crop with these factors in mind, it soon becomes clear that an understanding of potential growth locations must be obtained. What are the plant's moisture requirements, what sort of soil will it need, what temperature range? What about slope, surface stones, semiarid climates? These factors, and many others of a similar nature, must be plotted in relation to what is already known about the plant being considered. The result is a map of the United States divided into numbered sections that indicate where the new crop is likely to do well. If the sections don't interfere too much with the boss crops of the day — like corn and wheat — then a first step has been taken. The journey continues.

The factors, however, become more complicated. If a crop that can be grown on "unoccupied" land requires a heavy capital investment in exotic machinery, storage facilities, and intricate processing techniques, then it is not as likely to make a place for itself. Capital intensive farming is done on the most productive land; much of that is already taken. On the other hand, if a new crop's capital needs are modest, and its prepare-plant-harvest-process profile is, instead, labor-intensive, then another plus is notched. The marginal lands of the nation, and the world, are generally inhabited by lower-income people who need jobs as well as better nutrition.

In the United States, unlike some other nations, the growing and harvesting are among the less complex steps on the way from field to mouth. In this economic system, everyone along the way must make a profit to stay in business. So, no matter how marginal the land a new crop needs, no matter how many new jobs will be created in the newly plowed fields, if the farmer, the person who owns and manages those fields, can't sell his crop for more than it cost him to produce it, he is not likely to plant that crop more than once. Making the same mistake twice in farming is as hazardous as it is in other endeavors.

The crop, then, must be sold to distributors at prices that bring the farmer a satisfactory net return; it must also be acceptably priced so it will be purchased by consumers once it reaches the supermarkets of the land. There must be, as the agricultural establishment puts it, agronomic or horticultural feasibility combined with economic feasibility within the production-marketing-consumption system. In addition, any successful new crop must be compatible with the existing legal and institutional framework; it must fit within the requirements of land-use controls and acreage allotments, and within the capabilities of marketing and transportation networks, commodity interest groups, and research and agricultural extension services.

A new crop won't get very far if it damages the environment in any way: causing soil erosions, producing sediments that are transported by wind or water, promoting the loss of natural vegetation cover that has helped preserve important natural species — these are some of the ways a crop can hurt its natural surroundings.

And then there is the matter of the crop's social desirability. Does it create any health problems? Or will it, by the nature of its agricultural requirements, tend to change the scale of farming in the region? An existing system of small farms, for example, could be absorbed by a single large owner if a new crop made large-scale operations feasible. Such a sudden metamorphosis could result in profound social stress.

The journey no longer appears easy, does it? Our new crop, once started on the complex voyage from "there to here," must meet a number of tests and pass each of them, even though some of them may seem to have little to do with the world's nutritional needs, or the risks currently inherent in mankind's dependence on just a handful of food plants.

When SALUT set out to generate that list of 54 candidates to make the journey, it started with a list of approximately one thousand crops developed by the Plant Taxonomy Laboratory of the United States Agricultural Research Service in Beltsville, Maryland. That list was originally developed on a global basis and included many strictly tropical plants not suitable for the nation's temperate climate. Those warm weather varieties were dropped, as were those species that had no value as a food or fiber crop.

More than nine hundred plants were eliminated. The list of 90 crops which remains was circulated among more than one hundred persons throughout the 48 states (exclusive of Alaska and Hawaii), each of them an expert at universities, state agricultural experiment stations, United States Department of Agriculture research stations, and other organizations and agencies. After their replies (more than 60) were received and analyzed, the list was culled to the names of 76 plants; 22 of those were eliminated because their basic virtues related to the contribution they could make to industry, rather than nutrition.

The crops on the final list, according to the SALUT report, are adapted to greater or smaller segments of the 48 temperate-zone states. They can also be raised on marginal land, and provide a product critical to the larger environmental and energy challenges facing society. Each of the crops chosen has been grown and investigated by one or more crop experts in the United States. The list is the result of the combined judgments of some of the best-informed horticulturists and agronomists throughout the nation.

Why, you may ask, has some of your tax money been used to conduct such a thorough, complex, and extensive search? SALUT has given us its answer to that question, and, in a word, the answer is "diversity." There is a need for diversity in these days of monoculture and specialization. The Industrial Age pendulum has swung too far toward centralized, high-energy, mass-production agriculture. Diversification beyond the species currently mass-produced can result in the enrichment of available germ plasm, in new products, insurance against environmental or pathological catastrophe; it can mean greater land-use flexibility, energy conservation, and a means of expanding the small farmer's income base.

A National Academy of Science publication (1975) points out that one of every ten plants is either already extinct, or in danger of extinction. This amounts to more than 20 thousand species, many of which are potential candidates for use as food, fiber, or as a source of chemicals. And some have uses which may not even be guessed within conventional frames of reference. Early in 1979 a plant was "discovered" growing in Mexico that could change the way corn is grown around the world.

The plant, teocinte, may be the earliest ancestor of corn and, when hybridized with existing varieties can convert the crop from an annual to a perennial, thus lowering one of corn's major costs — the time spent plowing and seeding it each year.

Since the time of those Mexicans of nine thousand years ago in the caves near the port of Veracruz, mankind has used about three thousand plant species for food. Now we have reduced that number to less than 20. The end result of this accelerating trend to monoculture has been the creation of a unique set of problems. Among them are new susceptibility to blight, like the corn blight of the 1960s which came close to becoming a nationwide crop failure. Disaster was avoided by making use of blight-resistant strains taken from the germ plasm bank, but there is growing concern over the likelihood that similar actions will succeed in the future. The pressures of dwindling water supplies, ever more costly energy, more resistant pests, environmental pollution, and urban sprawl have created an ever more complex array of problems.

Oil — or rather, the growing constraints on its global supply — is another reason why new crops are needed. Petroleum-derived fertilizer, petroleum-derived pesticides, and refined petroleum products to fuel the gas-gulping legions of farm machinery — these are the foundation of contemporary agriculture. Yet, with our oil-import policies fluctuating severely in accord with political fluctuations in the Middle East, with oil exploration ever more expensive because it now takes place in the hostile environments of the Arctic and/or the open sea, the oil that sustains the large, monocultural producers of corn and wheat and soybeans is becoming painfully expensive. Farmers have driven their tractors to Washington to protest the shrinking profits their petro-based systems have brought, yet there is little Washington can do to create more oil, to lower oil prices, or to further subsidize increasingly expensive farm policies.

It is those high-energy policies that need to be changed, and some of the new crops may be able to help. It becomes a question of market access. If a new crop does solve some problems on the farm because it can be grown on more marginal land with a lower energy input and on a smaller, more labor-intensive scale, no good will come of it unless there is a market. What's grown for food must, eventually, be demanded as food. And, if a new crop is to feed the rest of the world, it should preferably be eaten in the world's richer countries first. One of the problems with some agricultural experiments in Third and Fourth World nations has been the tendency to offer crops that have no popularity in developed countries. Third and Fourth World consumers imagine they are being offered products richer folks have already rejected. That is not the way to persuade people to adopt a crop that may be "good for them." They've got to want to eat it first.

The same is true for housewives in Kalamazoo and Dallas: they have to think that what they find on supermarket shelves is more than just good for them; they want to believe that they are eating what the rest of America also thinks is "good," and there's a difference. Part of that difference has to do with market access. If there is a large supply of a given crop on store shelves, then the assumption that it's "good" generally follows. Example: bread made from whole grain wheat flour makes tasty bread that is nutritionally better; yet most American shoppers still buy bread made from bleached and refined flour because they think it's "good." That public taste is largely controlled by market access; there is a great deal more bread made from refined flour on the shelves. Whole grain flour has made advances over the years, but they have been slow indeed. Imagine the problems products made from a new crop could encounter at this level.

On the other hand, public tastes have a way of shifting. If a new crop can coincide favorably with those shifts, its chances for market access are considerably improved. Thirty years ago, for example, health food stores in America were all but non-existent, except possibly in Southern California. Today, they are everywhere and represent a multimillion dollar business that is still growing. Granola today is a household word, and people are sprouting mung beans, soybeans, and alfalfa. This sort of specialty market could become an oblique way of meeting the market access challenges a new crop must clear.

Some of those challenges are more readily apparent than others, and the list is too long to be itemized here. But among the more fundamental is the challenge posed by the need for new machinery. Cotton was never "King Cotton" in the American South until the invention of the cotton gin. If a new crop requires the invention of new devices and new machinery, then it is a crop that will not stay ahead of the pack when final choices are made and governmental priorities set.

Which raises another, less visible challenge. Like it or not, the United States government and the United States agricultural system are so closely interlocked that no basic change will take place on the nation's farms until that change also takes place in Washington: in the Department of Agriculture, in the department's research agencies, its consultants, its experimental crop units; and in the Congress and the White House where farm policy and farm price supports and farm subsidies are established. Most established crops already have their defenders, their lobbyists. Tobacco, for example, is a crop with defenders able to dip into a protobacco fund of millions of dollars to finance "educational" campaigns that will help the public see tobacco at its best, and help Congress and the White House keep tobacco subsidies at their best and tobacco research programs at their most extensive.

No matter how well a new boy on the block may fit into the environmental and energy neighborhoods, no matter what the potential market for his products, the people in Washington who make their living by protecting the vested interests of the old boys are not going to give in easily. The way they see it, there is only so much money in the government's agricultural till; if some is spent for new crop research and promotion, then less will be spent for wheat subsidies and tobacco research.

This reality causes some strange patterns to develop back on the farm. New crops are often introduced with an accompanying buildup and enthusiastic claims about their money-making potential. Farmers, typically eager to utilize land to its most effective money-making capacity, go all out with the latest plant being pushed by the Department of Agriculture, or the local extension agent. At the same time, however, defenders of established crops reassert their claims to subsidies, price supports, and market access. The experimental and hopeful farmer who risked his acreage on first-time crops looks at the end-of-the-harvest results and often learns that while the crop did indeed yield as well, or better, than advertised, and was, in fact, economical to grow and environmentally compatible with the land, it could not compete on a net-gain basis with its older rivals and the support they get in Washington and the marketplace.

The following year, the chastened farmer abandons the new crop and returns to the old standbys. The pattern is well known in agricultural circles. Efforts like the well-financed and long-term campaign to get Maine farmers to raise sugar beets in Aroostook County are typical. Hundreds of millions of dollars were spent there to give potato growers a profitable second crop; a huge sugar beet refining plant was built in the wake of years of soil testing which indicated sugar beets would do well in Aroostook County soil, and free seed was given to any farmer willing to plant it. For a while, it looked as if the success line had been crossed. Many farmers were growing beets and the refinery geared up to do the processing. With a sugar squeeze in the marketplace pinching tighter and tighter in the wake of Cuba's boycott on sugar shipments to the United States, prices held firm; on paper, the plan appeared highly profitable. In the potato fields, however, the reality appeared dismal. After their early enthusiasm had worn off, Maine farmers agreed almost unanimously not to take a chance on beets. They had little faith in the sugar beet entrepreneurs, they had a natural resistance to change, and they knew that if they stuck with potatoes, they would make something — perhaps not much, but something.

Beetless years rolled by, one after the other. Today the

refinery is silent, abandoned, the rusting skeletons of its machinery rattle in winds that blow topsoil from Aroostook's potato fields. If there were an epitaph inscribed on one of the vacant metal storage sheds, it might read: "You can lead a farmer to a new crop, but you can't make him plant it." In the final analysis, it is the farmer who holds the key. Once a volume crop is produced, marketing devices — advertising, packaging, and publicity — can persuade shoppers to buy it.

Knowing that, SALUT consultants produced an outline of the principle issues that must be discussed and resolved if any new crop is to survive and eventually thrive as a public commodity. That list reads like this:

I. Individual Entrepreneur Benefits
  A. Characteristics of the production-marketing-consumption system
    1. Producers
       Farm type, number and distribution of producers, convertibility of present farmers to new crop, convertibility of equipment and machinery, replacement of existing crops, etc.
    2. Marketing and processing
       Processing and storage requirements, transportation, convertibility of present facilities and institutions, technical problems and research needs, institutional arrangements, number and type of marketing channels, etc.
    3. Product and consumers
       Type of product, current and potential demand, type of consumer, number, price, volume, etc.
  B. Potential economic returns
    1. Estimate of land costs, production costs, sale price, marketing costs, etc.
    2. Estimate of marketing system costs
    3. Estimate of net returns
  C. Preservation of land resource

D. Energy conservation
E. Adequacy of institutions
1. Farm organizations
2. Commodity organizations
3. Research structure
4. Farm supply structure
5. Marketing channels
Collection, storage, and transportation organizations; processing and distribution
6. Advertising and promotional media
7 Governmental bodies
F. Legal constraints
1. Land use controls and subsidies
2. Taxation
3. Other constraints
II. General Public Good
A. Economic considerations
1. Economic growth
2. Stability, flexibility, and efficiency of production
3. Labor utilization
4. Import substitution
B. Environmental considerations
1. Land resource conservation
2. Hydrology and sediments
3. Wildlife and scenic qualities
4. Pest habitat
C. Social considerations
1. Social utility of product
2. Labor requirements
3. Influence on size and structure of farms

Fitted to that complex matrix, the grain amaranth is one of five finalists of the original 90 recommended for a major new-crop effort. One cannot help but wonder how different the amaranth's status might be if Cortez hadn't ordered those tens of thousands of amaranth acres destroyed more than four centuries ago. It seems logical to assume that if he hadn't the amaranth would be an old boy today, instead of a new boy trying to earn a place for itself in the established agricultural neighborhood.

Just how much work remains to be done before that place is earned is summed up in the final comments of the Soil and Land Use Technology report:

> *Grain amaranth, a potential specialty health food, has a good chance to generate an economic return to the producer and serve the general public good because it is adapted to a short growing season on marginal lands with cool temperatures and saline soils. It has a low energy requirement and a high quality protein content rich in amino acids, lysine, and methionine, which most grains lack. It can utilize excess rural labor provided harvesting can be done economically with hand tools or small equipment. In the crop development system, it occupies Step One. Agronomic research has not progressed in the United States and neither market studies nor feasibility studies have been carried out.*

Remembering that it took 25 years for soybeans to become an established and valuable crop, anyone convinced of the amaranth's great potential for providing the world with a new and productive protein resource can't help but wish Cortez had not been so efficient at his techniques of crop destruction. We are, it seems, having to begin all over again to get from "there to here," and, in the eyes of some observers (like the consultants at SALUT), Step One is just beginning.

On that score, however, as any visitor to the Organic Gardening and Farming Research Center can tell you, those consultants may be unduly pessimistic. Records at the farm for the past several years, and others gathered from around the nation and the world, indicate that amaranth "agronomic research" has indeed been carried out, and, in the process, a great deal has been learned.

CHAPTER SIX

Some of that research began in 1973 when Dr. John Robson contacted Bob Rodale. As the director of the nutrition program and a professor of nutrition at the University of Michigan, Dr. Robson had already begun a search for facts about the amaranth and other high-protein plants that had been nutritional staples of past civilizations.

Wrote Dr. Robson in a letter to Rodale on July 2, 1973:

> *In our present state of knowledge it is apparent that an abundance of descriptive material is available in the literature describing plant foods consumed by man in the past and present. Although there is much information relative to the proximate composition of these food materials, there are serious gaps in our knowledge of the nutrient content of foods eaten by man. The data suggest that the huge plant resources of the world provide a significant amount of protein and energy for its population, but there is a serious lack of information on the composition of these food materials as consumed. This is particularly true with respect to the biological value of the protein. In addition to the need for improved methodologies in measuring amino acids in foods, there is a notable lack of data obtained from animal feeding experiments.*
>
> *These deficiencies in knowledge would seem to justify a systematic program of research. First, there should be a search of the literature to document plant food materials with a potential nutritional value for Man. A further study of anthropological literature should be conducted to identify plant foods used by Man in the past and present, and the methods used to prepare them for consumption. Second, a search of the literature should be made to ascertain which foods have been chemically analyzed. Third, a proximate analysis of those foods on which data are lacking should be made both in the raw state and as consumed. Foods with a high protein content should be further appraised by simulated peptic and pancreatic digestion* in vitro, *amino acid analyses, and by animal feed-*

ing experiments. Finally, food identified as being suitable for consumption by man should be evaluated from a botanical and economic standpoint to see if commercial exploitation would be feasible. Such a program would have high expectations of augmenting the present food supplies of man and help to insure those of the future.

That letter, written five years before the SALUT consultants suggested that amaranth research rested at Step One, was the catalyst which sent Rodale scientists to Mexico, which brought the world's largest collection of amaranth germ plasm to the Research Center, and which set several other scientists and Dick Harwood to work learning as much as they could about the new-old crop. As it turns out, even a brief look at some of the results of that work should indicate how much more is known about the amaranth today than was known five years ago.

To start with the plant's beginnings, a good deal more has been learned about the seed from which it grows; and it's reasonably safe to assume that the two people in the nation who have learned the most are Laurie Feine and Charles "Skip" Kauffman, the two Research Center scientists.

Working in their laboratory-offices at the New Farm, in the greenhouse behind the main farmhouse, and in the fields that stretch from there back to the Pennsylvania hills, the two young, patient, and seemingly tireless researchers have tested, planted, pollinated, photographed, recorded, and measured hundreds of amaranth varieties. In addition, they have established genetic data which make it clear that the amaranth is a plant that has the sort of built-in genetic strength and diversity which will make it a dependable commercial crop.

The entire Rodale collection of more than three hundred varieties of amaranth seed has been planted and documented by Skip and Laurie, including rare types from Nepal, India, Hong Kong, and Taiwan. Morphological types have been classified, information about each plant has been recorded and

CHAPTER SIX

Charles "Skip" Kauffman, a plant breeder at the Organic Gardening and Farming Research Center, with an unidentified species that is a "Nepal grain type." It is an outstanding individual and a good parent plant for breeding because of its large, single plant head.

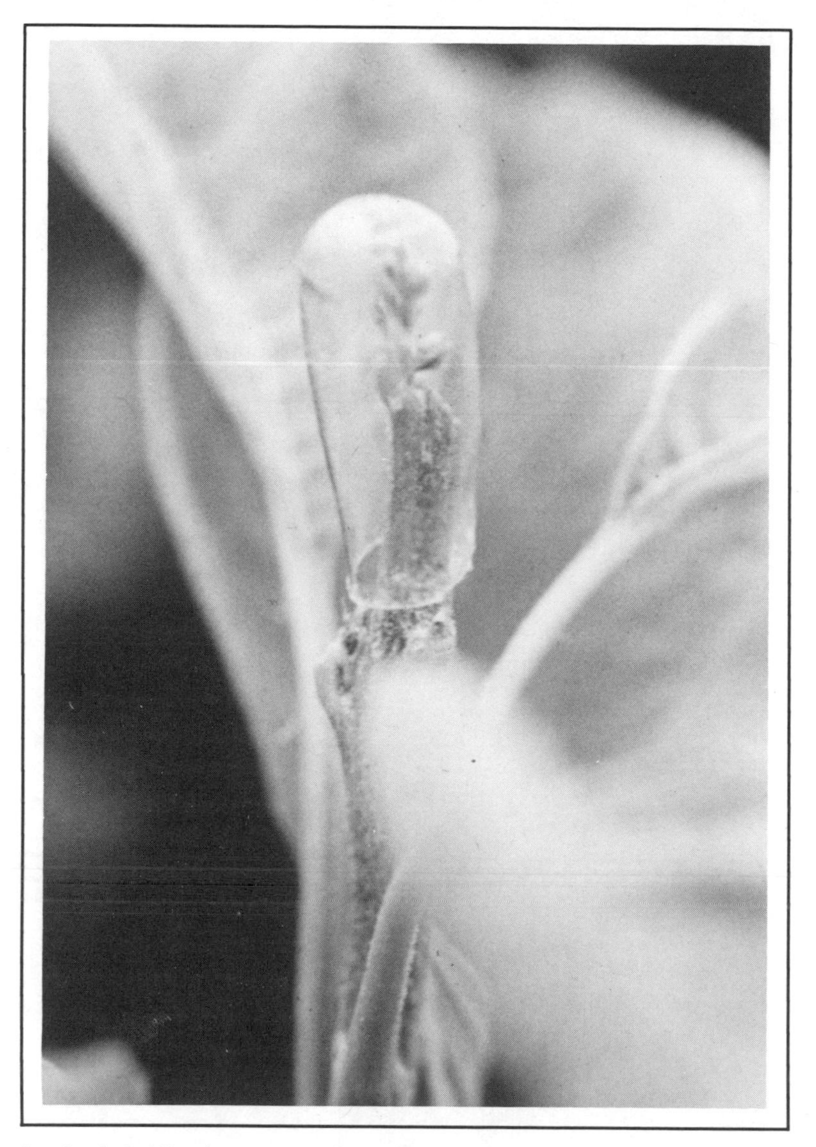

In the hybridization process the pollen-producing parts of the amaranth flower are removed, and a gelatin capsule protects what remains from accidentally cross-pollinating with another nearby plant before it can be hand-pollinated.

filed, new selections were made for testing and plant breeding studies. In addition, the plants were photographed, watched for insect and disease resistance, and harvested so more seeds could be added to the Rodale germ plasm bank as well as sent along for testing and study to the United States Department of Agriculture's Germplasm Resources Laboratory.

The results of the work and observations recorded every day fill five field books. If you want to know the source of any amaranth seed, or need a description of the plant's flower, stem, branching pattern, leaf color, overall shape, and other details, you can find it in the field books, as well as notes on insect and disease damage (if any) and the resistance of the plant to lodging — the scientific term for a plant's tendency to fail to stand erect, especially near harvest time when the seed heads are heavy and the plant becomes more vulnerable to strong winds and heavy rains.

Early in 1978, the two scientists began a painstaking effort to breed some of their plant collection in an effort to: (1) produce a variety of amaranth seed that would turn out to be dependable and would grow into a plant with a high grain yield, (2) have uniform growth characteristics (so it can be harvested with commercial, mechanized equipment), (3) be disease and pest resistant, and (4) grow a strong root system that would tend to prevent lodging, even though the grain heads grew heavy with seed.

The breeding process is delicate, close work. I watched Skip Kauffman on a summer day as he sat in a warm greenhouse, using tweezers and other miniature tools to duplicate the bee's gentle touch, taking bits of pollen from one section of one amaranth flower and transferring them to another, meanwhile making sure each of the other flowers were carefully covered to make certain there could be no accidental pollination.

For the layman, the process of plant breeding can be a mystery; I was fascinated by the notion that botanical science can "create" in a sense, a new variety. Yet, as Laurie and Skip

A. cruentus is bagged to prevent cross-pollination and assure pure seed.

explained it, the procedure is based largely on acute observation, common sense, and giving nature some help. Prior to their breeding work, the scientists had watched carefully each of the 308 amaranth varieties and selections from a large quantity of Mexican seed they had planted in the Rodale fields. As the plants sprouted, grew, matured, and reached their harvest stage, they were observed for their rate of growth, their eventual height, their yield — every characteristic that seemed most desirable for a plant that would become a farm crop.

When the growth process ended, 31 of the plants that had appeared to possess the best all-around qualities were marked and their seed planted in greenhouse starting pots. When those documented seeds produced a plant that flowered, Skip's controlled pollination procedures were begun — always in an effort to try to combine two amaranth plants that appeared to each have desirable characteristics that might be even more beneficial if they could be combined.

It was the second-generation seeds from those efforts that were later planted in the greenhouse for a test to determine how well the varieties had combined. As it turned out, Skip and Laurie discovered that the *Amaranthus hypochondriacus* (the direct descendant of the plant which supplied Montezuma with his annual tribute) responds well to crossbreeding, apparently does not tend to go into what geneticists call a depression — a reluctance of some plants to extend their genetic trademarks beyond one or two generations of crossbreeding. In 1979 these will be field-tested to see how they respond to natural climate.

Field trials testing yield potential were done using some of the parents of the breeding work. The yield champion produced an average of 1,886 kilos of grain per hectare (1,679 pounds per acre) of land; its closest competitor was more than one hundred kilos (220 pounds) behind. It showed scarcely any tendency to fall over, or lodge, even though some of its counterparts in other lots failed to weather winds which it had survived unscathed. Its rows were uniform in height, and when ready for harvest the plant had grown to about 1.5 meters (about 60 inches) — a dimension quite adaptable to existing mechanized harvesters. It is further adapted to mechanical

A. hypochondriacus, showing early seed head formation.

harvest because of its small, green flower heads borne at the top of the stalk. And, in addition to each of its other considerable virtues, it matured early.

If you go back and reread some of the characteristics which the SALUT consultants recommend for a new crop, you'll find that the 1978 grain amaranth yield champion fits, or exceeds, each of the criteria. But Laurie Feine and Skip Kauffman haven't thought of stopping with this variety. Oh, it will be rather heavily planted in the next growing season, and will find its way to other test plots around the country, but even as they are sprouting and growing, Skip and Laurie will be selecting for still better varieties tailored to fit the needs of different locations and different types of farming. The amaranth lends itself to such diversity — that's one of the several truths about the plant that Rodale researchers have already established, and it's a fact which would appear to take the crop's agronomic research a bit beyond Step One.

Late in 1978, Austin Campbell, chief of the New Crops Division of the United States Department of Agriculture, made a visit to the Research Center. He talked with Dick Harwood — the subject, amaranth. Campbell's wife, Judy Abbott, accompanied him; she's a horticulturist specializing in taste preference and nutrition. Along with his firsthand look at the amaranth and the place where so much agronomic research is being done, the USDA specialist left with 60 varieties of grain amaranth seed. In addition to taking root in the fields of Pennsylvania, the results of the work done at the Research Center will soon bloom under the interested eyes of the United States government's agricultural bureaucracy — the starting ground for any new crop that hopes to be taken seriously by Congress and the nation's farmers.

When it is, there shouldn't be any problem meeting the demand for seed. Based on calculations made from firsthand studies of high grain yielders, Dick Harwood has arrived at the following conclusions:

> *Amaranth seed weighs a little less than 1 gram (0.035 ounces) per thousand seeds. At a plant density of 20 thousand plants per hectare (around 9 thousand per acre) and with a 2 thousand kilogram per hectare (1,780 pounds per acre) yield (our potential here at present), the yield is 1 hundred grams per plant (100 thousand seeds).*
>
> *At a density of 20 thousand plants per hectare and assuming a five to one planting rate (a guess) it will take around 50 thousand seeds per acre. The seeds of one*

plant will plant 2 acres. One acre of seed production will plant 20 thousand acres. Fifty acres of seed would plant 1 million acres.

With minimal isolation amaranth is true-breeding.

If we sold a farmer five pounds of seed at $2.50 a pound ($12.50) and he increased it one time on 50 acres, the increase would plant 1 million acres.

This reduces seed cost to the farmer to practically nil. From that standpoint it is a perfect crop.

Perfection as a seed producer — and there is no reason to doubt the Harwood arithmetic, the proof is growing in the fields — does not mean perfection, or even performance, in other aspects of a plant's characteristics. In addition to meeting the needs of a farmer, his available land, his accustomed machinery, and in addition to capturing the attention of Congress and the entire agricultural establishment, the amaranth must also prove itself "good" as well as good for you.

Clean amaranth seeds (left) have been separated from the chaff (right).

CHAPTER SIX

Just how a plant works as a beneficial nutrient for humans is a process still not completely understood, not even by the experts. Bob Rodale has written that vegetarians report increased feelings of "wellness" and well-being, feelings he believes are caused by little-understood aspects of plant chemistry — a chemistry that interreacts with human chemistry to promote "good feelings":

> *Vegetables, fruits, seeds, and grain tend to be rich in fiber, but it is important not to think of that fiber in your system as merely doing the same thing as pure cellulose or sawdust. (One large, national bread company adds such purified sawdust to its bread to be able to make more fiber claims.) The fiber in fresh plant foods is now known to be active, in the sense that eating more of it can help to lower both cholesterol and blood pressure, and can insulate the system against toxic factors in other foods. In other words, the eating of more fresh foods from plants has introduced a whole new range of benefits into our diet that remains to be fully understood. Yet we are quite sure the benefits are real.*
>
> *We should also realize that plants harbor within themselves a tremendous chemical imagination. Each different type of plant is made up of its own specific panorama of compounds. And these chemicals, found in plants in such rich profusion, are not only the ordinary types — the kinds you would find in bottles on the shelves of a druggist or in a school chemistry laboratory. Many of them are so diverse, so unusual in their function, that even to this day plant chemicals are seen by scientists as a rich storehouse of inspiration. The leaves of unusual plants are still being analyzed to see whether the chemicals they contain might be useful as new drugs, as organic pesticides, or to suggest solutions to any one of dozens of chemical problems.*
>
> *Primitive peoples, like the ancient civilizations of*

Mexico, learned to use this chemical diversity of plants in many ways, but primarily to preserve their health. Thousands of years ago, people who knew nothing of chemistry identified those plants which contained basic drugs like quinine, curare, caffeine, and many others which are still in use today. They knew which plants were rich sources of vitamins, and sought them out. In their societies, human survival depended to a great extent on knowing how to put to use the chemicals in local plants.

I am convinced there is a benign connection between the plant world and the human world which we have yet to appreciate to its fullest extent.

If Charles Darwin were alive, it is quite likely he would be drawn into a stimulating dialogue by Bob Rodale's comments. As an evolving species, humans got their first nutrition from plants. And, as the earliest forebears of today's *Homo sapiens* survived on a vegetarian diet, their digestive systems, in turn, evolved over the eons to extract the most from the food headed their way. What developed was a sophisticated system for extracting certain protein-amino acid combinations that could then be converted to the protein the body needs. As the substance which is the basis for hormones, enzymes, energy, sexual vigor, connective tissue, and cerebral activity, the protein-amino acid combination which the digestive system designed itself for is the key to our development as intelligent beings.

As an ancient crop, the probability that amaranth was one of those plants to which human protein extraction is keyed is a strong probability indeed. Amaranth grain is still "food that is good for us," (Darwin would argue) because it is the food which our systems have evolved over the eons to utilize for our most pivotal nutrition needs.

Among the scientists working today to see just how well that nutritional theory — and others — holds up is Dr. Joseph Senft, a research chemist at the Organic Gardening and Farming Research Center. Among his other work, work that has been in progress for several years, is an on-going study of the nutritional qualities and specifics of the grain amaranth. In addition to his own research, Joe Senft also keeps in reasonably close touch with others in the nation, and the world, who are involved in similar projects. As of January, 1979, he was in touch with the following efforts (many of them supported by Rodale Research) to learn more about the amaranth's nutritional qualities:

—Protein and amino acid analysis of grain amaranth conducted at the USDA's Western Regional Research Laboratory;

—Dr. Robert Becker's work at the USDA's laboratory in Albany, California, where grain amaranth is being analyzed for carbohydrates;

—Dr. Peter Cheeke's feeding trials with rats at Oregon State University, where the animals are being fed varying amounts of grain amaranths;

—Charles Daloz, a graduate student of Dr. Henry Munger's of the Vegetable Crops Department at Cornell University is also conducting feeding trials;

—Oxalate levels in some grain amaranths are being determined by Dr. Der Marderosian at the College of Pharmacy and Science in Philadelphia;

—Dr. William Kendall of the USDA's Pastures Research Laboratory at Penn State is conducting palatibility studies of some grain amaranth varieties. Further studies of the nutritional qualities are also planned.

And, on a less technical, but equally important level of research, Joe Senft has been working to determine the precise dimensions of popped amaranth seed — popped almost exactly as it was by Mexican Indians centuries ago.

His report, which describes the light, airy results of that process in solid, scientific dimensions, reads as follows:

A comparison of equal volumes of popped and unpopped amaranth gave a weight ratio of 5 to 1. On this basis, popping increases the volume by 5 times (including air spaces). A similar comparison was made for popcorn. The ratio was 20 to 1. However, it is quite clear that the "packing" of the popped amaranth was greater than that of the popcorn. Thus as a volume expander for granola, flour, or mixed with other cereals (etc.), these ratios would probably be comparable (closer to the amaranth value) because the larger space between the popcorn would be filled by the smaller ingredients in such mixes.

And why, you might ask, would the possible effects of combining popped amaranth grain with granola be on Joe Senft's mind? To answer that question, you should keep in mind the complex process that's been outlined for getting from "there to here." One of the ways mentioned by SALUT consultants in their report to the National Science Foundation was popularization as a "health food." Surely, the grain amaranth

Dr. Joseph Senft, research chemist at the Organic Gardening and Farming Research Center, is in charge of extensive studies on nutritional qualities of the grain amaranth.

has those credentials, and just as surely, so does granola, the "health food" cereal that's now being produced as a breakfast food by some of the nation's largest food marketers.

Granola, as anyone who has tasted it knows, is "solid" food; a little goes a very long way. Market research experts — those people who make a science of knowing and anticipating popular tastes (and, in some cases, helping to create them) — have long thought that granola would be even more popular than it is if only it could be "lightened." After decades of eating boxed breakfast foods that added scarcely any weight to a tablespoon, many Americans were finding granola a bit too heavy.

Ah, but add a good measure of popped amaranth, and look at the result. Granola becomes instantly "lighter" because each spoonful holds some of the airy, popped grain as well as granola's traditional ingredients. And, in a nutritional bonus, the additive supplies additional protein to the consumer, as well as each of the other nutritional bonuses which have been detailed in earlier chapters. As a taste tester of the amaranth fortified granola, I can report that there's also an added nuttier sweetness — just a hint of it — within the finely textured, hull-free, white, popped amaranth grains.

Thus, yet another small step on the journey from there to here is begun. A respected and major marketer of health foods has heard about amaranth's potential and has contacted Rodale Press about the possibility of obtaining some seed for the testing. The answer to that query, as you might expect, was quick and positive. Bob Rodale wrote as 1979 began:

> *Our research work continues at an intensive pace and the results of plant breeding work and growing trials of this past summer are very encouraging. . . . At the rate we are now making progress, practical problems with the amaranth's culture should be solved very rapidly. Dick Harwood has told me that he feels that by the 1980 growing season the plant will be improved to the point where*

commercial production on a reasonable scale will be feasible.

We can send you immediately any quantity of seed you desire up to 50 pounds, from stock of varieties of grain amaranth we imported from Mexico. The quality of the seed should be satisfactory for preliminary work that you may want to do, although part of our program for amaranth development includes efforts to increase the size of the seed through selection and breeding, as well as other efforts to be sure the seed is of highest quality.

And so, yet another small piece of the whole, complex chain of events falls into place: there is the possibility of a small, health food application for the crop growing so vigorously in the Pennsylvania fields. There is, as anyone who has even a cursory understanding of that process now knows, still a "far piece" to be traveled.

For Dick Harwood, however, quite a distance has already been logged. He talked of what had been accomplished, there in that office on that cold January day:

During the year just past our research program has made real progress. Six staff members and one graduate student participated in the effort, three on a full-time basis. Grants were given by Rodale to five university programs to supplement our own studies.

We have initiated a comprehensive program to bring amaranth to the point of becoming an economic crop. That program includes the building of a scientific base for continuing development with such efforts as the comprehensive literature review and published bibliography, a world-wide germ plasm collection, nutritional analysis including several animal feeding studies, genetic studies, plant breeding, and horticultural studies.

Research progress has been most pronounced in the selecting of plant types which can be grown commer-

CHAPTER SIX

Organic Gardening and Farming Research Center director, Dr. Richard Harwood, with researcher Laurie Feine beside an experimental amaranth seed thresher.

cially. Varieties now available are uniform for height at 40 to 60 inches and can be combine-harvested. Trial commercial production will begin in 1979, with sufficient seed being available for planting of several thousand acres in 1980. The SALUT report concludes that amaranth has good overall market potential, with entry into the health food specialty market appearing to be promising.

Research backstopping in amaranth is gaining momentum. Breeding is well underway. The USDA at Beltsville, Maryland, is starting an improvement and selection program. Genetic studies are under way at the University of California at Davis and at Iowa State University. Nutritional studies are being conducted at Oregon State University, Pennsylvania State University, Cornell, the College of Pharmacy in Philadelphia, Muhlenberg College in Allentown, Pennsylvania, and here at the Research Center. Our test kitchen is developing amaranth-based food products.

Interest is high and growing steadily. The research programs now underway will give the crop impetus for several years and should bring it into limited commercial production. As a backyard garden crop, spread of grain types has been limited so far only by limited seed availability of improved types. It has received extensive coverage in the popular press. I feel that the crop can be quickly added as a base for high nutritional-value food products.

There is a contagion to enthusiasm, and excitement over the continuing saga of the amaranth's renaissance extends well beyond the walls of Dick Harwood's farmhouse office. Within days after he had impressed his visitors with his amaranth optimism, Harwood met with Bob Rodale for a further review of progress, and it was shortly after that meeting that Rodale wrote his editorial for the April 1979 issue of *Organic Gardening*. In a concise way, that editorial condenses the history of the

Rodale Press involvement with the plant, yet it also echoes from yet another perspective, the sense of accomplishment, the "something is about to happen" feeling that was so pervasive as 1979 began:

> *Excited by the possibility of finding a grain plant that could be grown easily and abundantly in gardens, I flew to Ann Arbor on a spring day back in 1974.*
>
> *"There's so much that can be done with the amaranth," Dr. John Robson told me as we toured his laboratories. "Getting more exact nutritional information is only the beginning. We don't even know if the plant will grow well in the United States, where the length of day is different during the plant's flowering period than it is in Mexico."*
>
> *Little did I know then that the trail of the grain amaranth would lead to the unfolding of dozens and dozens of other challenges. But we got started.* Organic Gardening *magazine gave Dr. Robson a grant of a few thousand dollars for his nutritional work. We also paid the costs of a mini-expedition to a remote part of Mexico to get amaranth seed. Within a few weeks of my trip to Ann Arbor, the first grain amaranth seeds were being planted on the farm that is now the Organic Gardening and Research Farming Center at Maxatawny.*
>
> *Yes, those plants grew well, flowered, and produced seed. They were beautiful, which we should have expected, because amaranths were adopted and selected for ornamental use soon after Columbus discovered them growing in the Caribbean during his first voyage. Some of the grain amaranths grew ten feet high. Others were shorter. There were color differences — some plants being a rich red and others green. I became enchanted by the amaranth. The plant had a beauty that put it in a class by itself. No food plant that I had ever seen before was so pleasant to look at as these newcomers to our farm.*

Soon our readers began to share our enthusiasm for amaranth. We ran several articles about the plant, and then offered packets of seed free to any person who would grow a small plot of amaranth, following careful instructions supplied by our Nancy Nickum Bailey, who I began to call the Amaranth Lady. By the second year of the project, Nancy had requests for amaranth seed from 14 thousand volunteer Reader-Researchers. Not all of them actually planted the seed or submitted a report as Nancy specified, but many thousands did. I'm sure that she and Dick Harwood were running one of the nation's largest horticultural research programs — at least in terms of number of participants.

Yes, most of the reseachers loved the amaranth — especially for the beauty of its large seed heads and luxurious leaves — but they reported problems, too. We were seeing the same troubles with the amaranth we grew at Maxatawny. Some of the plants grew so tall that, late in the season, they were blown over by storms. There was still too much wildness in the plant. In one plot you would see amaranths of all different sizes, colors, shapes of seed heads, and even with different branching patterns. Most important, we felt that because of its still-primitive nature, the grain amaranth plant was far from achieving the productiveness we felt it could attain. The amaranth as it then existed served well as a plant for the grain-gardens of rural Mexico, but we saw that a concerted development effort was needed to help it reach its true potential.

Then, beginning in 1977, something happened which quickly moved grain amaranth out of the realm of being a curious garden grain plant of special interest to organic gardeners and onto the center stage of interest in "new" plants that could contribute to the solution of world-wide food problems. Bear with me for a moment while I draw together the threads of this story. The word about amaranth's potential began to spread — all over the world, as a matter of fact. The National Academy of Science spotlighted grain amaranth; we held a seminar on amaranth for scientists, which was well attended. Most important, Dick Harwood traveled widely looking for more different types of grain amaranth.

Dick had worked for the Rockefeller Foundation for ten years as a specialist in the cropping systems used by small farmers in Third World countries. During 1977 and 1978, he completed consulting assignments in the Philippines, India, Nepal, and Central America. Everywhere he went in remote farming regions he looked for grain amaranths. And he found them — types which no one had known existed — because no one had looked for them in those places before.

He found a rich trove of new grain amaranths and new proof of its value in Nepal. In the far western part of that remote country is a farming region that is many days walk from the nearest road. A farmer living there who wants to trade crops for fuel or money must carry his harvest out on his own back. Some of the fields are 20 days walking distance from the road.

Amaranth grain, Dick saw, is carried much further distances than any other product, because the farmers had found that its value was higher. Maize they would carry only five days. Rice, being more valuable pound for pound, is carried seven days. But they will trek profitably for two weeks with amaranth.

On his own and with the aid of Nepalese scientists, Dick collected seed of varieties of grain amaranth which we had never seen before. This new germ plasm (which is the technical way of referring to plant material useful for breeding programs) let us see clearly that eventually the grain amaranth plant could be improved in a wide variety of ways. Not all of these seeds came from Nepal. Some were sent to us or collected in India, Africa, and other places. By the 1978 growing season we had 308 different amaranths growing at the Research Center in an amazing and beautiful display of plant diversity.

Some of the new plants were short, but with large seed heads. They will be useful to bring down the height of the plant. Other types grew straight, with no side branching. That is also a useful trait. By careful screening, the research technicians found some types with slightly larger seed. One of our major goals is to increase seed size.

Most exciting of all, our first efforts at an improved type (still using only the Mexican germ plasm) were shown last growing season to be remarkably successful. We now have a type of grain amaranth that grows uniformly to chest height, and has larger seed heads that

Seed heads of, from left to right: *A. hypochondriacus* (Mexican), an unidentified species referred to as a "Nepal grain type," *A. edulis,* and *A. cruentus.*

produce twice as much grain as the original semiwild plants Dr. Robson helped us get from Mexico. Agronomists who saw the improved amaranth growing last summer were impressed. We're sure that future breeding work, using the rich new germ plasm now available, will be even more successful.

By stressing what we have done with amaranth, I've perhaps given the impression that all work with the plant has been centered here. That is not true. We have made an important contribution by getting the germ plasm together, but much of that precious seed stock has been distributed free to other research centers and to universities where excellent work is being done.

The United States Department of Agriculture's effort can be a major factor in moving amaranth to where it will become an important crop in both gardens and on farms. But we understand that their commitment is for only four month's work, hardly enough to carry it through to a successful conclusion. Unfortunately, the department has only a small staff working on new crop development, while it still has many highly paid scientists occupied with studying and improving the tobacco plant.

Some gentle political persuasion is in order to help Washington see where its plant research priorities should be placed. I think now is the time to try to switch interest from tobacco to amaranth and similar underdeveloped food plants. If the many gardeners and small farmers who see in amaranth a very valuable new crop would write their Congressmen about this for help, we should at least be able to get an extension of the department's commitment to work with the plant.

There is one other important thing I want to say about amaranth. It relates to taste, and to uses for amaranth grain. Mexican Indians have for centuries popped amaranth grain, mixed it with honey or molasses, and made it into a highly nourishing, tasty, but not-too-sweet snack called alegría *— which means happiness in Spanish. We are sure that similar snack foods made from amaranth grain would be a big hit here among those millions of people looking for something good and natural to eat between meals. We have also found that popped amaranth grain, when mixed with granola, improves the flavor and nutritional value of that popular cereal. Amaranth granola has already been sampled by food company researchers and they're pushing to find large quantities of the grain so they can put that product on the market. Unfortunately, it will probably be another year or more before all the agricultural problems are worked out so that amaranth can be grown on a large scale as a grain. But the growing excitement of the food industry about amaranth grain vindicates Dr. Robson's early faith in the value of the plant.*

Of course, that must be an intrinsic presence if the journey from there to here is ever to be made; and there is no question now that after five years of scrutiny the amaranth researchers at Rodale and other places around the world are convinced of the value of the plant. It is, for all the reasons cited here, for all the criteria listed by new crop experts, a plant for the postindustrial age — an era we have already begun.

But eras do not begin and end in an orderly fashion; just the opposite is most often true — times of transition are uncertain times, often tumultuous and traumatic. Those who benefit most from the past resist the future with the stubbornness of a man clinging to a branch to keep from falling. And even though the amaranth has value nutritionally, even though it grows in marginal soils, has a root system that makes it easy on water and supplies and helps prevent soil erosion, and can be raised in drought-prone areas like those of southeast Texas, it will be grudgingly welcomed by large-scale farmers until that value has been proven beyond any shadow of doubt, until, in fact, the crop is built into the existing agricultural establishment.

Thus the journey from "there to here" is roundabout, long, and given to many side roads and turns that seem to be going nowhere. We already know that existing machinery — combines, planters, harvesters — can be modified to fit amaranth crops, but, ironically, probably won't be needed for years. They won't because the amaranth's beginnings as a major crop will come in tens of thousands of backyard gardens, in dozens of small, specialty farms, and as a market item meeting the needs of a special sort of attuned consumer. Only after that developmental stage has been successfully passed will the crop enter the process of becoming a staple.

The first step, however, on that journey has been taken. There are producers of foods for the attuned consumer interested in the grain amaranth's potential. There are scientists at work establishing its nutritional verities. And there are agronomists, and horticulturists interested in improving the stock. Perhaps more important, there are a great many Americans more conscious of the importance of what they eat and why. More and more supermarket shoppers are beginning to question the important differences between what is considered "good" in the lexicon of popular taste and what is good for us.

The end of that road is a long, long way out on the horizon of human comprehension. Just as we have crossed five centuries to rediscover the amaranth, we may also be about to cross eons of evolution to discover some truths about diet. As a genus which has come comparatively late to eating meat, as a species which evolved over the millennia from a gatherer of plants rather than a hunter of wild creatures, *Homo sapiens,* and especially Western Man who eats more meat per capita than any other counterpart around the world, may discover that plants have taken a dietary second place only at a high cost to our culture.

Theorize for a moment on the differences between plant gatherers and meat hunters. The first connotes a kind of gentleness, a society in which both male and female shared the basics of survival. Aggression was not a part of everyday living. With the advent of hunting, nearly every cultural characteristic was altered. Males became the hunters, the aggressors. And, because they had to overcome the fear that was often a part of the hunt, the hunters developed traits that had formerly been subdued: they became concerned about images of bravery, they developed a pride-of-violence; those who killed the largest and fiercest animals were those who became hero figures — different heroes from those who had been so benignly acknowledged when the gathering of plants was the primary survival activity.

With the conversion from plant harvester to creature killer came the traits that have stayed with Western Man ever since. They are the aggressive, belligerent, exploitive behavioral traits so personified by Cortez and his conquest of a people closer then to being harvesters than hunters. He who began with the conquest of the sabertooth continued to destroy the Aztecs and, almost incidentally, the amaranth which had been so important to the ancient harvesters of Central America.

It is possible, then, to make the case for a return to a more varied, more protein-rich vegetable diet on the grounds of becoming more civilized as well as more nutritionally fulfilled. The beef eaters of America are discovering quite rapidly that the habit is becoming increasingly costly as the price of the energy required to convert grass to hamburger continues to escalate with each tremor from the Middle East's oil fields. Indeed, beef and veal showed the highest percentage price increase of any supermarket "staple" during 1978, and there are no predictions that the trend will ameliorate. Instead, the predictions take the opposite tack: meat prices will get higher, and still higher.

At the same time, there is a more subtle change taking place: a tendency toward a less aggressive, less combative society — a society, if you will, with a growing constituency that would no longer sanction or demand the destruction of the

Aztecs and the burning of their fields. There is an ever larger number of Americans on the eve of postindustrialism who are questioning the values of violence and aggression.

They are the same individuals who show a new concern for investigating the interlocks between the psyche and the soma, the same people concerned with better nutrition, with growing instead of hunting, with investigating the historic links between amaranth and the metaphysical. And they are also the same Americans who will make the first scattered, small-scale, but critically important efforts to help the amaranth along on its journey from there to here . . . . the voyage from the rolling fields of an experimental research farm in Maxatawny, Pennsylvania, to the kitchen cabinets of nearly every home in the nation. As the record makes clear, even with the help of SALUT and the Rodale research team, that is a long and complex journey.

One factor which remains to be looked at, and which we have not previously discussed, is the state of the amaranth in other parts of the globe — places where the plant may never have been the staple it was in ancient Mexico, but where, on the other hand, it has been in moderate, steady use for centuries. There are places like that, and they, too, will play a role in getting this new-old crop from "there to here."

Chapter Seven

When the word "amaranth" is used in Africa, India, and parts of Latin America, the term can convey two quite different meanings. It depends on where it's used. In some parts of the same region of India, for example, discussion of the plant will be based on the assumption that it is a grain producer. Yet, perhaps in the next village over, on the other side of the mountain, there is little or no understanding of the amaranth's potential as a producer of a high-protein cereal crop. Instead, it is known as the source of one of the tastiest leaf vegetables in the native diet.

For the amaranth is one of those rare double-duty plants that supplies its growers with a grain crop and a leaf crop at the same time. There are, however, amaranth types better suited for the production of grain and others cultivated through the centuries for their leaves, which are picked when the plant is young and boiled gently for a short time, much in the way so many Americans cook their spinach.

The leaf amaranth has played a minor role thus far in this narrative of the new-old crop that could do so much for the world's nutritional and environmental needs; its time has come, however, and a detailed look at the "other" amaranth appears in the following chapter. The division around the globe is not quite as tidy. In Central Africa, parts of Mexico, South America, and in different regions of the Indian subcontinent, there is some overlap in the way a vast population of the world's farmers of small, subsistence farms utilize this hardy and surprisingly versatile plant. Unlike the American farmers who will view with some hesitation the arrival of a new boy on the agricultural block, farmers of the Third and Fourth Worlds are not as concerned about market access as they are about winning the life-long struggle against starvation. And that is surely one of the reasons why the amaranth has been cultivated somewhere on five continents for centuries.

Of those places, India currently has more amaranth varieties growing in more backyard gardens over a wider climatic range than any other region. In recent years, as United States interest in the plant has increased for the several reasons I have previously cited, a number of researchers have visited India with the single-minded purpose of learning more about the plant that somehow made the journey centuries ago from the valleys of Veracruz to the slopes of the Himalayas.

One of those researchers is Peter J. Claus, an associate professor of anthropology at California State University. With the assistance of graduate student Anida Weyl, Claus has compiled one of the most current and complete (albeit brief) reports on the variety of small, backyard farms and gardens where amaranth is being grown by Indians in many different villages and in many different climates. Among his observations of the cultivation of grain amaranths, Claus makes the following comments:

> *The cultivated amaranths have been identified in South Asia (India, Ceylon, and the countries of the Himalayas) since Linnaeus' time in the early eighteenth century. Commonly known as a garden flower, the stately, brightly colored prince's plume was easily and frequently noticed by South Asian travelers, especially in the isolated hill areas populated by tribal peoples. Some of the travelers may have confused these amaranths with another important South Asian food crop, the chenopods, since from a distance, they appear similar in silhouette. A*

further confusion arises because each community calls amaranth a different name and sometimes both chenopods and amaranths are similarly named.

Even the early travelers found amaranth growing as a staple grain. In the hills of northwest India where it is still known as bathu (batu), *the bread made from the seeds is considered to be a common food. It is usually grown along the edges of fields or in gardens.*

Amaranths are grown over a large area of the North Indian plain for use as potherbs in curry and the like, with little attention paid to the seeds. Near the Himalayan foothills, however, the opposite is true: only the seeds are utilized. When it is grown as a grain crop, as in parts of the Gujurat state in India, the yield is about three times greater than when it is grown as a scattered crop.

When amaranth is grown for seed, especially if it is to be marketed, care is taken to discard the black shiny seeds which occur as a variant of the cultivated species from time to time. Besides being ground into flour for use in breadmaking, the seeds can be cooked and used in a gruel which is the common practice in Nepal, where it is commonly called marchi *or* nana; *or popped and rolled in sugar or honey and made into round balls called* laddu.

Farther north, throughout the Himalayas, including Nepal and Bhutan, amaranth cultivation is widespread. It is often planted in unirrigated fields and seems to grow better at elevations exceeding 15 hundred meters (5,886 feet). Reportedly, it has been cultivated in areas of 25 hundred to 35 hundred meters (9,810 to 13,734 feet) — heights too high for human habitation. In these areas of glacial terrain, the ground is covered by about ten feet of snow for half the year or more. The plants are stunted in size, but nonetheless are grown, for few crops are as hardy as this one.

In Kashmir, amaranth is known as ganhar. *After*

special preparations, it is eaten by the Hindus of this area on fast days. A little farther to the south in the Kulu Valley of the Punjab, where about two thousand acres are planted annually, it is known locally as seol *or* seriara. *The grain is also known as* chuya *or* amardana *in the valley, while the salty gruel made from milk and the parched-then-boiled seeds is called* phambra.

In Mussoorie, chaulai, *as grain amaranth is known there, is one of the important crops of the monsoon season. It is grown above 18 hundred meters (7,063 feet) and is planted with millet, another important grain crop throughout South Asia, in poor, stony soil. It is said to produce in high quantity but is considered inferior in taste to other grains. In a village located in the Detra Dun district of Uttar Pradesh, Gerald Berreman reports that* chaulai *accounts for 20 percent of the grain harvest. There it is planted during the summer rainy season and harvested in September.*

Early travelers also reported the cultivation of amaranth in far southern India. Francis Hamilton-Buchanan, on his journey through India in 1800–1801, saw amaranth cultivated along with yams, millet, pulses, and sorghum, using a slash-and-burn technique. Although he fails to mention how the amaranth was used, it may be surmised that it could have been used as feed for chickens which he noted were in great abundance. A more recent report from the South Indian hill regions cited amaranth growing in fields with maize, beans, and squash — all of them thought to be New World crops. Presently it is ground and mixed with sweetened milk to make a porridge. It is also placed on funeral pyres, giving it a special ceremonial significance.

Amaranth is beginning to appear now in the Delhi market and throughout Madhya Pradesh as well as in Gujurat where it is known as rajgira. *It seems likely that the crop will spread and gain in economic significance.*

Today it is a substitute for the more costly white poppy seeds which are used in sweets and festival foods. Amaranth is a high-yield crop, able to grow in a variety of elevations and climates with relative success. In India today it is being intensely researched as an alternative food source.

In South India, I had the opportunity to observe amaranth being grown along the southwest coast during a field research expedition in 1975–76. There it appeared as a common kitchen garden crop, grown by nearly every peasant household. Three cultivated varieties are found. The native name is padpe (in the local Dravidian language, Tulu). A dwarf variety called dogoli padpe and two tall varieties of Amaranthus polygamous, identified as boldu (white) and kempu (red), are recognized. While the main staple crops of the region are rice and coconut, amaranth does have an important place among the vegetable crops and the peasant's diet. Most significantly, it has the potential to play an important role in future agricultural development of the region.

Amaranth is regarded as a potherb, the leaves and taproot being the parts for which they are grown. Although the author never saw or heard of the seed being grown for consumption, it would be surprising if these were not gathered and used for food, especially by the poorer families. The people of this immediate region appear to be unaware of the cultivation of A. hypochondricus as a grain crop, despite the fact that this crop has been grown in the nearby hill regions.

All three amaranths are grown in small garden plots along with yams, bitter gourd, finger gourd, beans, and okra. Only a dozen or so of each of these plants, including amaranth, are grown in each garden in what appears to the Westerner as a rather haphazard order. The main gardening season is during the dry months, October through May. The rainy months, June through September, are inundated with up to two hundred inches of *monsoon rain in which garden vegetables would be washed out. Amaranth seedlings for either eating or transplanting are available in the rural open-air peasant markets. Mature plants for cooking can be found there in season, too.*

It is certain that amaranths have been cultivated in this region for a long time: (1) it bears a distinct name; (2) it is a common vegetable, familiar to everyone; (3) although it is used sparingly, all parts of the plants are used in various ways, including medicinal uses; (4) it figures prominently in at least one folk legend of probably ancient native origin. It is curious then, why amaranths, particularly the grain amaranths, do not play a more prominent role in the agricultural economy and nutrition of the people of this region. This is a very important question. Population density of this region is one of the highest in India, with even a rural population of more than five hundred people per square mile. Although because of assured favorable rainfall each year, there has never been an absolute famine, there is the prevalence of severe malnutritional deficiencies — protein, vitamins, and minerals — and these are becoming worse by the year. Expanded amaranth cultivation could help to solve these nutritional problems.

What stands in the way? It may be that the climatic conditions during the season amaranth must be grown are not suitable for its proper maturation. Our survey of amaranth cultivation indicated that grain amaranths do best at higher elevations (one thousand meters [3,924 feet] and above). Further, it has been suggested (Singh, 1961) that the grain amaranth is sensitive to photo- and thermoperiodism. Since seeds would have to be sown during the winter months, yields may be very small. If this is so, special varieties would have to be developed for the specific local conditions before the plant would have a great impact on the agricultural economy here.

Even if varieties could be adapted to the local condi-

tions, we would still have to acknowledge the difficulties involved in getting people who are by tradition geared to rice cultivation and a rice-based diet to accept any other grain. Wheat flour and millet, both grains eaten in other parts of India, are available in local stores but are rarely purchased. Millet and maize can be successfully cultivated, but are not. To be sure, the parboiled rice traditionally eaten in the area is a reasonably healthy food. The present-day inadequacies are of a hidden nature. Today, more and more people are bringing their grain to the mill to be polished, losing a large part of the grain's nutrients. The pulses and coconut which have always complimented the rice diet are becoming proportionally more expensive. Development of short-stalked, high-yield rice has reduced the availability of fodder and consequently decreased the output of dairy products. It is a sad irony to see that attempts at modernization have actually contributed to the alarming increase of protein deficiency diseases.

The easiest solution would seem to be to expand the use of grain amaranth in ways which would not compete with rice. Even so, working amaranth into the people's diet is a complex consideration. Mixing amaranth seed flour in rice and rice products, as a few people do with millet, is one answer. But suggesting this would be like suggesting that Americans work soybeans into their hamburger — the smarter and/or poorer families would, but the majority of middle-class families would stick to their traditional diet, preferring to be unhealthy than change.

The early reports of travelers in South India give us a clue to a different approach. Where grain amaranth was grown in South India, we find an increase in the numbers of chickens raised. Each South Kanara household grows a dozen or so chickens per year, and this constitutes the major meat consumption. What primarily limits the households from growing and consuming chick-

ens and eggs is that the chickens must compete with the family for rice, the primary chicken feed as well as human food. There seems to be no obvious reason why amaranth grain could not be grown as chicken feed, thus increasing both egg and meat production. Chickens lay at their optimum when given a feed with 14 or more percent protein and high amounts of calcium and other minerals and vitamins. Amaranth grains contain more than this amount of protein. Furthermore, it is higher than all other grains, except millet, in both calcium and phosphorus. Cattle and water buffalo could be fed on whatever of the greens and stalky parts of the plants the humans do not consume. Should amaranth prove to be a good fodder crop, the net effect would be an increase in the protein consumption in the villages regardless of the food preferences which are likely to hinder the expansion of amaranth as a human food.

The logic at the base of a program focused on the premise that grain amaranths could become a new Indian poultry food, thereby raising the overall protein intake of many Indian villagers is quite a different approach to the conventional marketing process dynamics that have been considered in the United States. Nevertheless, the idea — like the amaranth itself — could make the journey, this time from India to the North American continent.

The poultry feed aspect of the amaranth, should it develop further in India, would have the support of that nation's several agricultural and botanical institutions, where interest in the plant has been slowly gaining momentum in recent years. Dr. T. N. Khoshoo, director of the National Botanical Gardens, and a leading scientist there, Dr. Mohinder Pal, are among those who have begun to develop larger libraries of amaranth research. They are at work, much as Skip Kauffman and Laurie Feine are in the Western hemisphere, on discovering and developing new strains of grain amaranths which will increase the plant's grain yield without detracting from the hardiness which allows it to grow successfully so high in the Himalayas and under such marginal conditions.

And, from the Indian Institute of Horticultural Research, the following tips on how to best cultivate amaranth were compiled by horticulturist P. J. Mathai:

> *Amaranths are short-day plants and hence when planted towards shorter days (winter months) they may bolt early. This is especially true in "chhoti chaulai" types. The best sowing time for vegetable amaranths in South India is August-September during the rainy season. The "badi chaulai" types are mostly direct-sown or transplanted. "Chhoti chaulai" types are mostly direct-sown. Grain types are usually transplanted. The seed rate is 1.5 to 2.0 kilograms per hectare (1.34 to 1.78 pounds per acre). Final spacing at 20 by 20 centimeters (about 8 by 8 inches) is observed to be optimum for vegetable types. This may be achieved either by thinning or by transplanting. In general, higher plant population per unit area is found to have better yield. Fertilizer response is reported mostly in case of nitrogen and then phosphorus. Higher nitrogen doses are reported to increase the protein content. General recommendation is 80 to 100 kilograms of nitrogen per hectare (71.2 to 89 pounds per acre). This should be applied in split doses. Not much work is available on the nutrient requirement. Organic manuring yields best results. The crop requires good moisture throughout its growth. The first cutting could be made 30 to 40 days after planting. Delay in cutting makes the plants fibrous but is reported to increase the iron content in the plants. Subsequent cuttings could be obtained at 8- to 10-day intervals. The types entirely grown for greens are reported to yield 15 thousand to 16 thousand kilograms greens per hectare (13,350 to 14,240 pounds per acre) (e.g., Co-1, a variety recommended by*

> *Tamil Nadu Agricultural University) and dual purpose types 12 thousand kilograms greens and 1 thousand kilograms grains per hectare (10,680 pounds of greens and 890 pounds of grains per acre).*

Which, it would seem, may well be good advice for farmers and gardeners everywhere. As many observers of the international scene know, the popularity and acceptability of transnational foods is often more affected by social rather than nutritional considerations. Thus the popularity of a plant in America is likely to have as much influence on helping to spread amaranth cultivation in India as any set of helpful planting hints from the Horticultural Institute. This basic, but frequently overlooked reality, was well stated recently by a botanist, Dr. Subodh K. Jain of the Department of Agronomy and Range Science at the University of California at Davis — on sabbatical leave in India for germ plasm collection.

After a trip of nearly five thousand miles through his nation by car and train in search of amaranth information, Dr. Jain reported that he found many grain varieties growing along the rivers and in terraced fields of northern India. "Simply beautiful" is how the botanist described the colorful crops which seemed to be considered quite ordinary by the rural farmers who grew them. Taking note of such details as the rather uniform farm policy of tossing away any black amaranth seed found mixed with the white, and using only organic fertilizer in their fields, Dr. Jain also discovered fewer amaranth plots as he left the northern hills and traveled through the lower plains — where the crop would logically appear to be able to produce higher yields.

"I asked one of the plains farmers about why he did not plant more grain amaranth," reports Dr. Jain, "and he replied, 'If Americans grow it, then we will find it more popular here too. I promise we will.' "

Some forms of foreign aid, it would appear, may depend more on what is done by Americans at home than on what

Dr. Subodh Jain, well known for his germ plasm collections, moved amaranth research significantly forward with the large amaranth germ plasm collection he brought back to the United States from northern India. His student, Holly Hauptli, has done extensive work in amaranth genetics and agronomy at the University of California at Davis.

United States goods are shipped to ports around the world. Another form of foreign aid, it would also appear, has more to do with what we can learn from other cultures than with what we consider best for them.

Just as the notion that amaranth may also play a role as a

supplemental poultry feed originated with the Peter Claus visit to India, so too has an insight into a new-old way to raise amaranth resulted from the visit of Daniel K. Early to a small village in Mexico.

As they do in India, Central American small-scale farmers continue to grow amaranth in much the same ways as their long-ago ancestors did. Dan Early, a professor of anthropology at Central Oregon Community College, traveled to the Tuyehualco region of Mexico, near Mexico City, and on that journey gathered the amaranth information which he shared recently at the annual meeting of the American Anthropological Association in Los Angeles, California.

In a process that in 1978 echoed the techniques used by the Aztec farmers of 1478, Early discovered yet another method of starting, transplanting, cultivating, and eating grain amaranth. Notes from his journal tell the story:

> *One focus of recent appropriate technology research has been the search for high-protein plant foods which can be cultivated on a small scale without the high energy inputs of the hybrids promoted by advocates of the Green Revolution. Amaranth, because of its high protein content, delicious taste, high yields, and environmental adaptability is one such plant.*
>
> *Perhaps amaranth's greatest asset, however, is its taste. Unlike soybeans whose flavor must be disguised to be made palatable, amaranth tastes delicious. In fact it survives in Mexico as* alegría, *"joy" candy.*
>
> *As a processor of solar energy, amaranth ranks among the small group of C4 plants such as sorghum, millet, and maize, which most efficiently utilize sunlight and carbon. C4 plants grow faster and use less water than the more common C3 plants. Surprisingly adaptable, amaranth thrives in Mexico's tropical Morelos as well as severe Colorado. Another plus is that amaranth seems to be relatively drought-resistant.*

Dr. Daniel Early knows Mexico well, for he has spent much time there, seeking out areas where amaranth is grown and where *alegría*, the popped amaranth grain candy, is made.

Its cultivation and preparation techniques used by native Mexicans are well suited to intensive gardening and small-scale farming applications.

In the Tuyehualco area near Mexico City, the Aztec's chinampa *"floating garden" system is still being used to grow amaranth seedlings which are then transplanted to highland fields.*

This ingenious horticulture method suggests ways of combining high-protein grain production with intensive gardening/aquaculture systems such as those pioneered in the United States by the New Alchemy Institute and others.

The chinampas *are an ancient system, combining aquaculture and intensive horticulture. Originally built* in the swampland surrounding Mexico City as a land-reclamation project, these chinampas provided much of the food base for the Aztec civilization. A highly efficient production system, some of these plots have been producing continuously for more than two thousand years. Much of this ingenious system can be adapted for appropriate technology applications.

The chinampas or "floating gardens" are not actually floating. They were constructed by scooping mud from the irrigation canals and piling it up between post and vine walls. Layers of water weeds, which serve as compost, form the upper section of the plots. These weeds are covered with a layer of fresh mud, which provides a fertile medium for planting. The intricate system of canals surrounding the chinampa plots contain carp and other edible fish. The canals also provide organic fertilizer and mulch in the form of abundant green algae, which is skimmed off the water and applied to the plots.

One of the most interesting aspects of the chinampa system in terms of appropriate technology is the ingenious method of seedbed preparation. The seedbeds are formed from canal bottom mud, which is cut into little squares before it is completely dry. Seeds are dropped into these squares of dried mud. When time to transplant arrives, each square easily breaks from the other like a little peat pot, holding the roots intact. Everything the villagers grow, except corn, is started in these seedbeds and then transplanted to another location within the chinampa plot or, as in the case of amaranth, transplanted to outlying fields.

To make his amaranth seedbed, the chinampero first excavates a plot, about 6 feet wide, 25 to 30 feet long, and 1½ inches deep in the chinampa alongside the canal. Then, standing in his canoe, he scoops the rich canal bottom mud into his boat with a long-handled shovel. All the sticks and foreign matter are separated out and tossed back into the canal. Then, using two buckets, he passes the mud to a partner on shore who pours it into the excavated seedbed.

The following day, when the mud begins to dry, the grower cuts it into little squares in which he drops the seeds. To cut the squares, he uses a special rake called a cuchilla with ten sharp knife blades 8 inches long, spaced $1\frac{1}{2}$ inches apart. The rake is custom-made by Mexico City blacksmiths, and costs about seven or eight dollars, minus the handle. First he makes a series of cuts across the width of the bed, reaching out to the opposite side and pulling the rake towards him. He then places a plank on top of the mud to distribute his weight and works his way down the bed, making lengthwise cuts, forming squares about $1\frac{1}{4}$ by $1\frac{1}{4}$ inches. After he cuts the squares, he cuts a small stick, about $\frac{3}{8}$ inch wide and 2 inches long, from a nearby tree. It's used to poke a hole about $\frac{3}{8}$ inch deep in the middle of each square, although some people just use their fingers.

He drops a pinch of amaranth seeds, roughly eight, between the thumb and first finger into each hole. He then tosses a layer of dry cow manure over the bed, covering the seed holes. After a couple of days, he removes the manure. He takes the larger pieces off by hand, and sweeps off the rest with an improvised whisk broom, made by joining together leafy tree switches. This leaves a fine powder of manure filling the holes and covering the seeds. The grower then leaves the amaranth alone to germinate.

There is generally enough moisture in the mud of the seedbed, so that it doesn't need to be watered for the next five days. He then waters the seedlings every two days with the algae-rich water of the canal to maintain constant humid conditions. Of the seeds planted in the beds, three to six will germinate.

When the plants are six to eight inches tall, which is usually around 20 days after sowing, they are ready to be transplanted to outlying fields. The villagers usually

Near Mexico City amaranth seedlings are grown in *chinampas* (floating gardens) made from rich mud scooped up from the bottom of nearby canals.

The mud is left to partially dry and is then formed into "peat pots" with a rake made just for this purpose.

Six to eight amaranth seeds are dropped into the large hole in each of the "peat pots," and they are then covered with dry manure to keep them moist and protect them from the hot sun.

A few days later, the larger pieces of manure are swept away, leaving a fine powder to fertilize the young plants.

The seedlings are watered every few days with algae-rich canal water until they are large enough to be transplanted in the fields.

transplant around the eighth of June to take advantage of the summer rains. First, the grower lifts the mud squares from the seedbed and easily separates them from each other without injuring the root system. There are three to six seedlings in each square, and their roots keep the squares from crumbling. This way there is no problem of tangled or broken roots.

The grower transports the plants to the highland fields in a square carrying box, a huacal, *dating from pre-Hispanic times. The sides of the* huacal *are formed by lashing sticks laid on top of one another log cabin style to form a square. The bottom is made of flat wooden slats, covered with a layer of grass in order to fill the spaces between them and provide a cushion for the plants. He lays the seedlings in overlapping horizontal layers, first pointing in one direction, then in the opposite. Finally, he covers the top seedlings with another layer of grass. This protects the tender seedlings, and also helps to retain moisture.*

He then secures the box with a rope and carries it on his back to highland fields with the pre-Hispanic trum line. In the fields, he plows rows, curcos, *roughly one meter (about 39 inches) apart and about 30 centimeters (about one foot) high with a horse- or mule-drawn plow. In general the plots seem to be under one hectare in size. Field size, however, remains a subject for further study. Following the plow, a sower carries a bundle of amaranth squares on a* maguey, *agave cactus branch, in one arm and tosses the plants one square at a time between the rows about one meter apart. Each square contains three to six young amaranth plants. The plants are not sown closer than one meter because "one takes strength from the other." Following the sower, another person actually transplants the seedlings. He scoops out a small hole, places the plant square in the hole, and with another movement pulls soil down with both hands from the rows and packs it around the plant for support. The three to six plants within each square are not thinned but allowed to grow together.*

Twenty days after transplanting the amaranth is fertilized. Some growers use the "GuanoMex 10-1," chemical fertilizer sold by the government. Others use organic methods. If cow manure is used, it is distributed in mounds, roughly 20 centimeters (8 inches) in diameter by 8 centimeters (3 inches) high, around each plant.

Villagers using chicken manure sprinkle a circle roughly ⅜ inch wide, ¾ inch high, and 3 inches away from the plant, because chicken manure "is strong and burns the plant." After applying fertilizer, the grower plows again, covering roughly half of the 16-inch plant with soil to support the long amaranth stalk.

About 30 days after transplanting, when the plant is roughly three feet tall, the grower returns to his fields for a final cultivation. He plows about 16 inches of earth round the base of each plant, and removes any weeds that have sprung up. If the villagers discover wild amaranth (pigweed) near the fields, they pull it up so the wild plant won't hybridize with the domestic amaranth. They often boil, then fry and eat the tender leaves of the young wild amaranth. In this cold region, flowering usually occurs about four months after sowing.

Villagers in the Tuyehualco area also sometimes intercrop amaranth with bush beans, Frijol negro. *The beans are sown between the amaranth rows, spaced the normal distance of three feet apart. Tomatoes and chili peppers are also intercropped, but here the amaranth squares are sown twice as far apart as normal (six feet) with tomatoes or chilies planted in between.*

When amaranth reaches a height of six feet, usually three months after the flowering, the plants are ready for harvesting. In the Tuyehualco area, the amaranth harvest begins around November 20 and continues into early December. The villagers use sharp sickles to cut the plants close to the ground. If the seeds have dried on the stalks, they can be winnowed the same day. If not, they lay the stalks in the field rows to dry for two or three days. Once dry, they separate the seeds from the stalks by beating and screening through a sieve made from a wire screen, the kind used in construction to screen sand. They nail wire mesh to a rectangular wooden frame with four legs, raising it about four feet off the ground. In place of the wire screen, a few growers use an ayate, *a cargo cloth of loosely woven maguey fibers. They stretch the* ayate *between four poles stuck in the ground, and place the amaranth stalks in the* ayate, *then beat them with brooms, made from the resilient switches of the perlilla plant.*

The villagers place a white cotton cloth under the

screen to catch the seeds as they fall through. They then screen the seeds a second time through a finer fly mesh screen. The seeds are sometimes dried in the sun. The growers spread them out on a large straw mat, and turn the seeds over by "dancing" on them, moving the seeds with their feet so they will dry evenly.

Once dried, the amaranth seeds are stored in cotton sacks; villagers prefer sugar sacks. To keep out the humidity, they often place cotton sacks inside large plastic feed sacks. Villagers say that if you store the seeds this way, they will keep for up to ten years. Furthermore, rats and other pests do not seem to bother the seeds. This ease of home storage is another advantage of amaranth.

Villagers pile up and abandon the dried stalks. Formerly some people used the stalks for cooking fuel, but this is no longer done. Growers mix corn stalks with cow manure in compost piles, but do not use amaranth stalks because of the slow rate of decomposition of their tougher fibers. The stubs are plowed under, and the field is left to rest or is planted in another crop. Amaranth is generally not sown twice in the same field, because it is thought to deplete the soil.

Villagers measure yields by volume, and an average yield in the Tuyehualco region is 4 cargas *per hectare or 4½ bushels per acre. In the Morelos area with a more temperate subtropical climate, growers obtain yields almost twice as large. Due to the lack of water, Morelos growers do not practice* chinampa *cultivation and transplanting. They sow amaranth directly in the fields.*

Not only does the chinampa *system provide an example of an intensive aquaculture/agriculture combination, it also illustrates how amaranth can successfully be transplanted. The homestead gardener in areas where the growing season is short should be able to sow substantial grain crops, starting the amaranth seedlings in greenhouses or cold frames and transplanting the seedlings to his fields.*

CHAPTER SEVEN

From India we learn that grain amaranths can provide a supplemental poultry food, and from a small village in Mexico we learn a new-old way of organic cultivation as well as an amaranth recipe for curing stomach troubles.

As if to emphasize the totality of their restoration to a free and independent democracy — some four centuries after the Conquest — the Mexican government has initiated a large-scale program of amaranth agricultural reseach. In a project scheduled for completion at the end of 1979, the Center for Economic and Social Studies for Third World Nations in Mexico City is working in conjunction with Mexico's Department of Agriculture to classify the different varieties of amaranth flour, to further study germ plasm classification with the goal of improving the seed. In addition, Mexican scientists and agriculturists are looking for ways to improve harvesting and planting techniques. Part of the effort involves building contacts with every other Latin American nation working in any way with amaranth cultivation and/or research.

Based on what little published material is available, the search for amaranth information in South America appears likely to become a long one. As Jonathan Sauer has reported, the species that has grown most widely among the Andes and other South American locations is *A. caudatus* — a plant used for grain found in the cool and temperate valleys of Peru, Bolivia, and Argentina. As I have noted previously — and with some wonder — the way the plant is utilized, its preparation as a ceremonial food, and its metaphysical overtones each have remarkable parallels with the amaranth in Central America and Asia. How these stunning similarities spread across the continents is still a mystery, and one which may well come into clearer focus as the Mexican government — now in communication with Rodale Research — pursues its Latin American efforts.

They will have to look closely in the backyard gardens of remote Andean villages, for it is only there that the amaranth is grown on any sort of regular schedule. Even then, it is what agronomists call a "relic crop" — one grown more for traditional reasons, handed down over the centuries, than for commerce and nutrition.

In general, it is grown at elevations of six thousand to ten thousand feet, although it is known to grow successfully down to sea level. In the Andes, amaranth is identified especially with Huanca people, and can be found planted in the barrancas of Huancavelica, Junin, Ayacucho, and Apurimac. Just as the natives of the Himalayas do, the farmers of the Andes make certain they plant only the white seeds — with one exception: in Ecuador a dye amaranth is grown and its red pigment, called *sangorache* by the Indians, is used as a ceremonial food coloring. Perhaps, if the trend toward sanctions against chemical food coloring in the United States continues to gain momentum, the *sangorache* amaranth of Ecuador will also be "rediscovered." One of the red chemical food colorings used most widely in the United States is, after all, known technically as "Amaranth #2."

From Ecuador a "natural" and potentially nutritional dye, from Mexico an organic gardening system, from India a new insight into poultry feeds — from North America, Central America, South America, and Asia flow trickles of information about the amaranth. They are trickles on their way to becoming streams, and soon those streams will become rivers. There is, however, one continent still to be heard from: Africa. There the emphasis shifts from the amaranth as a producer of grain to a plant that has been used for centuries as a potherb and a vegetable staple. Thus, the time has come to introduce the amaranth's "other half" — and some say the better half. Of all the world's crops, old or new, this is one of the very few that can supply both grain and potherb. Why Africans have all but ignored the grain potential and stressed the leaf harvest is yet another of the vast continent's many mysteries.

There is, nevertheless, a considerable amount of information on the uses of the amaranth leaf as a vegetable staple, and it is information which reveals something of African agriculture as well as the plant that has been growing there for thousands of years.

Chapter Eight

—Green is beautiful.

—If no sufficient milk is available to an infant with few teeth, give it mashed leaf vegetables with its first foods.

—Give at least 50 grams or one handful of leaves a day to a toddler.

—Let school-age children and adolescents have liberal quantities (100 grams a day or two handsful) of leaf vegetables as long as they are growing.

—The same for pregnant women and nursing mothers because they eat for two.

—If the staple food is very starchy, or if no meat or fish is available, increase the quantity of leaf vegetables in the daily fare.

—For beautiful eyes and strong teeth, give your child leaf vegetables.

—Let your leafy greens only cook for the shortest time possible; cover the pan.

—Use the opportunities you have, rely on local resources, and do not readily believe the claims of importers of expensive foreign products.

—There is no situation in the inhabited tropical world where you cannot have leaf vegetables with or without serious efforts. Usually, however, without.

Taken from a booklet on tropical leaf vegetables published by Holland's Department of Agricultural Research in Amsterdam, that advice — intended for all nutritional consultants and educators in the world's tropical zones — includes the essential truths about leaf vegetables, one of the more overlooked and underrated foods on the globe.

Leaves are every plant's factory division; without leaves there is no growth. Their main function is to recycle and convert the carbon dioxide of the air into carbohydrates. They are the base for the great variety of compounds we know as protoplasm — the vital component of cells. It is the composition of protoplasm that is ultimately responsible for the nutritional values of leaves.

Thus, leaf proteins represent the largest store of all food proteins in the world. The annual yield of protein per hectare from a leafy crop is larger than from any other farming system. Yet, in spite of every effort to solve the world's nutritional problems, the leaf has yet to attract much attention from the agencies in nearly every nation that is searching for ways to help close the malnourishment gap in Third and Fourth World countries.

In addition to protein, leaf vegetables contain calcium and iron, and the vitamins carotene, thiamine, riboflavin, niacin, ascorbic acid, and folic acid — each of them essential to good health, eyesight, strong teeth, and proper growth. And, of all the leaf vegetables grown in the world's warmer latitudes, none is more tasty or more widely used than the leaf amaranth.

In the villages of Southern Dahomey on Africa's west coast, an almost daily routine centers around the plots of vegetable amaranth grown in the backyard gardens of the villagers. Served as a supplement to two of the three daily meals eaten by every village family, the boiled leaf amaranth dishes echo similar dishes served in the warm and humid latitudes between the Tropic of Cancer and the Tropic of Capricorn the world around. Central America, India, Indonesia, and above all, every nation in Central Africa, utilizes the leaf amaranth as a staple in home diets, and as a market vegetable of considerable economic importance.

CHAPTER EIGHT

The numerous vernacular names of the leaf amaranth underscore its wide distribution: Chinese spinach, Sudan amaranth, African spinach, Indian spinach, Ceylon spinach . . . the list goes on. And there is a reason why English-speaking people have converted amaranth to their more familiar "spinach." The number one leaf vegetable of the tropics is almost always prepared and consumed in much the same way as the green vegetable Popeye has helped make famous.

In tropical Asia, the various species of leaf amaranth are — as they are in Africa — the most widely grown leaf vegetable. Indonesia, the Philippines, Taiwan, Burma, India, China, and New Guinea, where rows of green amaranth bloom in platforms filled with marsh mud and built to adjoin the native houses, in each of these places the plant is a familiar staple. African nations like Senegal, Ghana, Zaire, Nigeria, Uganda, Ethiopia, Mozambique, and others would be shorn of their number one home-garden crop if the leaf amaranth were taken from them.

The three primary species that make up the crop on each of the continents are *Amaranthus cruentus, A. dubius,* and *A. tricolor.* In one of the most detailed studies compiled of the plants and their place in the cultures of the tropics, the author, Dr. G. J. H. Grubben — a botanist with the Royal Tropical Institute in Amsterdam — describes in fine detail the planting, cultivation, harvest, and cooking of the leaf amaranth. Dr. Grubben's treatise — "The Cultivation of the Amaranth as a Tropical Leaf Vegetable" — is based primarily in the small settlements of Southern Dahomey where he traveled to study the plant and the people who depend on it. He stayed in West African villages through an entire growing cycle, and his photographs and painstaking descriptions must be one of the more thorough leaf amaranth annuals on record.

"It is," said the eighteenth century botanist G. E. Rumphius, "one kind of the potherbs. One won't find a house so small with a bit of land without also encountering a small bed of amaranth." Herr Rumphius was writing about Indonesia when he tucked that note in his journal, but his message is much the same as the one Dr. Grubben communicates about the amaranth in West Africa.

There are, in his photographs, neatly planted beds of *A. cruentus*. The straight rows of small, leafy plants are tended by villagers who skillfully thin the fast-growing annuals. A good planter can poke a hole in the ground with his finger and transplant as many as 60 seedlings in a minute. The young plants are taken from their seedbeds early in the morning and replanted before the heat of the midday sun. Growing in their organically fertilized plots, watered daily by the children of the village, the plants are ready to be harvested in three weeks. After the first harvest, which involves cutting off most of the leaves, the amaranth regenerates in another two or three weeks. Many times, over a period of three months from the initial seedling transplant, the villagers can take four leaf harvests before the amaranths are uprooted and another planting series begun when the first generation has gone to seed. If there are more plants growing than the family can eat, then freshly cut and washed leaves are taken to the market in Porto-Novo and sold.

Like all types of amaranth (and only in a few Ethiopian villages is any grain amaranth grown in all Africa) *A. cruentus* produces abundant seed. Village farmers have only to harvest and clean a small portion of it to obtain the source of their next, new crop. For tens of millions of people in the world's tropical zones, the dependable availability of the green leaf vegetable is the only source of vitamin A — the substance that growing youngsters must have if they are to enjoy proper vision and — in many cases — prevent blindness. Much of the Third and Fourth World population is able to see only because the amaranth is so prolific and can produce such a bountiful leaf crop on such a small bit of land. No wonder that on the mangrove coasts of West Irian, where villages are built on piles sunk in the salt sea bed, local Papuans often use discarded canoes filled with laboriously collected humus as their floating amaranth gardens.

Indeed, when one looks at the list of the amaranth's proven advantages and benefits, and considers the plant's productivity, one can only wonder about why the plant is not a more familiar green in the United States. If such a list were compiled, it would almost certainly include the following benefits:

—Ranks among the most efficient crops in production of dry matter, protein, and vitamins for each day that it occupies a given area of land.

—Contains protein of high nutritional quality which supplments the amino acids found in a diet high in cereals and legumes.

—Has a high content of minerals important in human nutrition.

—Tastes as good as spinach and is preferred to spinach by many people.

—Adapts to high temperatures when many of the common leafy green vegetables such as spinach, cabbage, and lettuce are unproductive.

—Is harvestable as a whole plant when 20 to 40 days old, or tips of branches picked repeatedly when plants are older.

—Harvested young, most of the above-ground part of the plant is edible.

According to horticulturists Henry Munger and James Deutsch (who prepared a paper on the plant for Rodale Research) the leaf amaranth is one of the most efficient of all plants when it comes to converting solar energy into human food. Yields as high as 13 tons per acre of fresh material (30 metric tons per hectare) have been reported for only 30 days of growing time in Taiwan. Few crops grow as quickly, and its brief occupancy of the land needs to be taken into account when its productivity is compared with other crops.

Munger and Deutsch report that protein makes up 25 to 30 percent of the edible dry matter in amaranth greens. This puts it in the same range as most legume seeds except for soybeans. Of course, neither greens nor legume seeds are eaten in dry form, so it may be more pertinent to compare them when cooked. The Food and Nutrition Research Center of the Philippines repots 4.4 percent protein for boiled amaranth as compared with 2.2 percent for boiled rice and 7 to 11 percent for various dry beans eaten after soaking and boiling. Amaranth is not greatly different from most other leafy greens in the protein percentage of its dry matter, but when fresh or boiled, it is higher in protein than the others because it has a higher dry matter content. This means that an average serving of amaranth greens contributes more protein toward the daily requirement than most other leafy vegetables — almost twice as much, for example, as lettuce and spinach. A one hundred-gram (3.5-ounce) serving of amaranth would provide from 7 to 10 percent of the daily protein needed by an individual, varying with age and other factors influencing the requirement. People have been known to eat two hundred to three hundred grams (7 to 10.5 ounces) of amaranth per day, in which case the protein contribution would be quite significant.

Another way of looking at protein is the rate at which it is produced since this is an important factor in determining its cost. The experiments in various tropical countries cited previously for dry matter production indicate that amaranth produces from about 4 to 18 pounds of protein per acre per day. This puts it in the same league with soybeans as a protein producer, since a United States Department of Agriculture report shows that highest average yields of soybeans are producing about 5 pounds of protein per acre per day.

Leaf proteins may have value in a diet in excess of their percentage contribution since they frequently contain amino acids that are not adequate in other components of the diet largely made up of cereals and legumes.

Protein, vitamins, and they taste good too. Authors who have written about the leaf amaranth describe that taste as: ". . . a very good vegetable. . . . surpasses spinach and swiss chard. . . . one of the most delicious vegetables. . . . the best of all tropical spinaches in both flavor and food value." Reading comments like this, and perusing the scientific reports on the plant's unique place in so much of the world's diet, one cannot help but wonder why amaranth greens are not a larger part of the American diet. To help learn more about the reasons, 28 varieties were planted during the summer of 1976 at the Research Center. When they were harvested, cooked, and tasted by a number of typical American diners, the results of the project were subsequently reported by *Organic Gardening*'s planning editor, Jack Ruttle:

> *Amaranth makes an excellent green vegetable that ought to be more commonly grown. It's a greens crop, we learned, that won't go to seed until late in the summer. And hot weather doesn't affect the flavor of the leaves. Amaranth is even fairly drought-tolerant, whereas most greens available in seed catalogs are cool-weather crops. So it is surprising that amaranth hasn't become a summertime favorite, especially since one of the choicest Asian varieties,* A. gangeticus, *has been available under its Indian name,* tampala, *for more than 30 years from a major seed company.*
>
> *All the kinds we planted are used as greens somewhere in Central and South America where amaranth began as a major grain crop.*
>
> *All but two kinds thrived in our Pennsylvania summer. And their flavors seemed nearly the same to the unfamiliar palates of our taste testers, although one or two stood out as more tender or bitter than the others. The opinions at our dining room generally agree. Only a few of all those people liked amaranth as fresh salad greens. Opinions seemed related, naturally, to what other greens folks ate. Fans of endive and young beet tops usually liked amaranth shoots in salads too.*
>
> *Cooked, however, most of the varieties we tried were hits. Our home economist, Anita Hirsch, fixed our amaranth greens steamed, in soups, lasagna, many ways. Both tasters and researchers agree that it is a good green. A lot*

Anita Hirsch, director of Rodale's Experimental Kitchen since 1974, has created and tested hundreds of amaranth recipes and has developed techniques for grinding, toasting, popping, and sprouting the grain.

of them ranked it with their favorites. The taste is mild and all its own — most like spinach, not at all like beets, chard, or the cole family. Several testers also praised it as easy to clean because leaves are smooth and borne well off the ground.

Mark Schwartz, our food technologist, sent samples to an independent laboratory to be analyzed for certain nutrients. We wanted to compare it to our traditional greens. The results showed it better in two key nutrients, calcium and vitamin C, than spinach, chard, and beet greens. Not much of the calcium in those greens is available to our bodies anyway, since high oxalic acid levels tie it up. But the analyses said the amaranths have less than half the oxalic acid content of spinach. Amaranth greens are also high in iron and vitamin A.

One of the taste testers when the various amaranth greens were served in Pennsylvania that summer was Bob Rodale, the man who had sparked the initial interest in the plant when he responded to Dr. John Robson's phone call two years before the first leaf amaranth dishes were served to the Rodale staff in the company dining rooms. For Bob Rodale, the enthusiastic reports on the ease of cultivation and general palatability of the plant he had helped to sponsor must have been encouraging. Two years before he had editorialized on the possible advantages of his favorite plant, stressing then not so much the leaf amaranth's good taste as its value to general health and well being:

This particular plant, like many semiwild foods, has an important advantage in that it supplies the fiber and bulk which our intestinal tracts need so badly. Modern nutritionists ignore fiber in food because it is not only indigestible, but doesn't have any nutrients to which we can give a name. How, then, can something that is just a filler be essential to health? Well, we have found that it is very important, and perhaps essential.

Dr. D. P. Burkitt, an English physician who has worked in Africa for many years, noticed that native peoples who lived on rustic, wild diets containing much fiber never get appendicitis or diverticulitis, a disease in which hernias form in the intestines. They also hardly ever get cancer of the colon, which has been increasing among people in industrialized countries and is now the second most common form of cancer, after lung cancer. People who naturally eat a lot of fiber in food also seldom get gallstones, varicose veins, hemorrhoids, and pulmonary embolisms caused by deep vein thrombosis. They are uniquely protected from heart attack, and, most interesting of all, are seldom fat. Hunger seems to be easier to control when a person's innards are liberally supplied with fibrous bulk.

Vegetarians naturally have more bulk in their diet than people eating the typical American diet, because meat is almost totally lacking in fiber. Sugar likewise has no fiber, and white flour is a poor source indeed. You can firm up this nutritional link with our wild past by eating less meat and sugar, consuming more vegetables and fruits, and learning to make whole grains a part of your regular diet. Bran is one of the best of all sources of fiber in the diet. You can buy raw bran in a health food store, add a few teaspoons of it to your cereal or other foods, and gain another important connection with the advantages of primitive living. This so-called "bran hypothesis" has been gaining headway rapidly in medical circles, and I'm sure that we are going to see major changes soon in the typical diet as a result. But you can gain those advantages now without waiting for others to awaken to an obvious fact.

Evidence seems to arrive daily supporting the contention that organic farmers have had for years: unprocessed food is superior to the processed varieties. In a recent issue of Ecology of Food and Nutrition, *D. H.*

Calloway and his coworkers report that the traditional Hopi and Papago Indian foods were consistently higher in essential minerals, especially calcium and iron, than the federal commodity foods donated to the tribes.

All the above factual information dealing with the quality of highly processed foods points us in one direction — to the homestead and the cultivation of semiwild animals and plants. As a further incentive it should be noted that the amaranth is perfectly adapted to growing in small homestead gardens, and needs no complicated equipment for harvesting. (As a matter of fact, some amaranth authorities, such as Edgar Anderson, believe that the plant will grow anywhere.) Allow me to illustrate. If you want to free yourself from the annual 15 percent (or possibly greater) rise in food cost each year, you can grow corn, wheat, or oats if you have the land, but these crops require equipment for threshing, shelling, and processing.

On the other hand, semiwild plants like amaranth do not require that kind of postharvest attention. They are not a problem because, when these plants were used widely, no one had more than the most simple kind of equipment.

I fully expect that future homesteaders will again be using these "forgotten" plants to create a new and valuable source of stored food. We will be doing that not to ward off starvation, the threat that faced our ancestors, but to arm ourselves against the twin modern enemies that are almost as devastating. I am referring to inflation and to the diseases of our insides caused by too much eating of spineless, devitalized modern food.

It was shortly after that editorial appeared that Bob Rodale and his associates at Rodale Press began one of the most unique experimental efforts ever attempted in American agriculture. Instead of waiting for those "future homesteaders" to learn more about amaranth on their own, Rodale executives made a decision to use their publications to offer free amaranth seed to anyone who asked for it. In return, although there were no promises extracted or contracts signed, the home gardeners, farmers, homesteaders, window box tenders, amateur and professional horticulturists, and anyone else who made requests were mailed their seeds and asked to report back to Rodale Press with the results of their first work with the amaranth.

What Rodale created, in effect, was a new approach to getting a new crop from there to here. Instead of waiting for the popularization of the amaranth through established systems, Rodale used a publishing distribution mechanism to reach growers who otherwise might never have heard of the amaranth. The United States Department of Agriculture, the major seed firms, the professional tastemakers were bypassed. Instead, the message went directly to the people, much as it had over the centuries in Mexico as the word about the new crop spread from village to village, culture to culture, and from continent to continent — before the printing press, before radio, before the telephone or television. In response to that Rodale message, people began growing, harvesting, preparing, and eating both grain and greens amaranth in nearly every state in the nation. Some of their responses, you may be interested to know, are a part of this book. You have only to turn the page.

Chapter Nine

"The idea," explains Bob Rodale, "was to involve the ultimate users of the new garden and farm crops in experimenting with and evaluating amaranth on the grass-roots level. We needed a record of their experiences throughout the country's many climatic zones. We wanted to know how the amaranth would perform in all kinds of weather — rain in some places, drought in others. And we needed more information about how the crop really works. What's the best way to thresh the grain, to mill it? What about lodging, seed shattering, uneven growth, and maturation. We needed some answers, and I couldn't think of any better place to look than in the home gardens and farms of America.

"So we put a small notice in the magazine (*Organic Gardening and Farming*) early in 1975, and what we came to call the Reader-Research project was begun."

That first year, three hundred volunteers responded to the request that they plant, grow, and harvest amaranth, and keep accurate records on every phase of the work. During the growing season of 1978, three years later, the Reader-Research corps numbered 13,500. Over the three years, amaranth, on one scale or another, had been grown in 25 thousand American gardens. As volunteers, each of the test growers added enthusiasm and zeal to the usual budget of patience and hard work that's required to get the most from any crop.

Along with the harvests at home, Rodale harvested letters,

Nancy Nickum Bailey heads up the Amaranth Reader-Research Program at Rodalc Press, one of the nation's largest horticultural research programs, in terms of number of participants.

telephone calls, sketches of devices to thresh amaranth seed, to dry it, and photographs of proud, or not so proud, researchers standing by their progeny, or bending over it in a symbolic search for the problems that had kept their first crop from living up to the researcher's dreams. Coordinating, deciphering, and responding to every amaranth communication that reached the Rodale offices in Emmaus, Pennsylvania, was Rodale staffer Nancy Nickum Bailey. She heard it all — what the gardeners and farmers liked and didn't like about the amaranth. She heard tales of freezes, droughts, dust storms, floods, dry spells, rampaging wildlife, pests, and pestilence. She listened to complaints about uneven ripening, about plants that seemed to break, not bend, in wind storms, and about seed that shattered and spilled from the plant heads just as it seemed a bumper crop was about to be harvested.

From every major climatic zone in the country, Nancy Bailey has collected three years worth of reports. That "grass-roots" file, along with the other amaranth research material she has found and organized since 1974 put her in possession of more amaranth information in one place than anyone else in the world (that Nancy knows of). Among the objectives of the Reader-Reseach effort was the assembly of more data on the plant's United States geographical limits, yield, optimum planting density, and how the plant would grow in nonirrigated plots. The Amaranth Lady, as Nancy has been titled, now has that data, and more, including the general observation that the amaranth is a hardy plant, adapted to a wide range of latitudes and climates across North America.

In addition to planting, cultivating, and harvesting, the Reader-Researchers were asked to record temperatures during the entire growing season, to keep track of rainfall, to record the time of seed germination and seed harvest, to weigh the harvested plants, and to thresh the seeds and weigh them accurately enough so fairly precise yield totals could be obtained.

The quality of the seed milled from Maxatawny over the three years, however, improved with each season, and should continue to do so as Skip Kauffman, Laurie Feine, and others like them make steady progress in the effort to produce an amaranth seed that can be depended on to germinate and to grow into a strong plant with a high yield. But no matter what sort of amaranth the Reader-Research corps planted, no matter what the restrictions on watering — and those were strict, generally insisting on no water other than natural moisture — and no matter the emphasis on no chemical fertilizers or pesticides, the amaranths planted in those thousands of plots got more tender loving care than your average plant.

Perhaps that is what was responsible for the high yields, because the interest the volunteers took in their test subjects was certainly apparent in the letters that arrived on Nancy Bailey's desk. Not only did the reports include the basic information that had been sought, but there were suggestions about everything from the phases of the moon (for optimum planting times) to a variety of methods for threshing and grinding the amaranth grain. Each of the suggestions was evaluated, and from them have come pilot models of low-energy threshers and flour mills — two home appliances all but forgotten in the American kitchen as well as on the American farm where even the farmer who grows wheat and corn is generally many miles from the mass production mills where the kernels are threshed and ground.

There was more, of course, in the letters. There were human stories of effort and triumph and disappointments in each of the reports that came from the farms and gardens to the Amaranth Lady's desk. Read excerpts from some of those letters for yourself, and you can share with Nancy Bailey and the Reader-Researchers yet another facet of the many-faceted amaranth adventure.

From David Cavagnaro of Santa Rosa, California:

Last April (close on to a year ago!) you sent us seeds of mitla beans, a variety of other pre-Columbian beans and squash, and Mexican amaranth. We are finally ready to send in the results. I must hasten to apologize for the long delay. This has been our first year on our own new 180-acre homestead, and the chores have kept us behind in processing the harvest and finishing the records until now. The number of projects we had on our mind this last growing season will also account for our rather homespun science, such as the missing weight for an amaranth plant which the sheep ate when they broke into the barn where the plants were drying, and the absence of certain mitla records, such as flowering dates. All told, however, our trials were a success, and we deeply appreciate having had an opportunity to try these early varieties.

In spite of the sheep, the amaranth experiment was a roaring success, though we got off to a late start with it because we were behind in the new garden. We ate the thinnings, and found them of the finest quality. The plants which remained grew to astounding proportions, were spectacular to look at, and produced a bounteous crop. We are sold on it as a grain crop; it exceeds our beans by quite a bit in pounds per acre. It is definitely a plant for our area.

From Russell Millsap of Woodland, California:

During World War II I served as a weather forecaster in the Air Force and spent one year in eastern China and one year in northeastern India. Consequently, I am thoroughly familiar with those areas in both countries where amaranth grows. It appeared that the amaranth in those countries grows at a fairly high level, anywhere from two thousand feet on up. However, both areas are subject to a monsoon season which brings fairly humid and wet summer climate.

I was able to save quite a bit of the seed after plant-

ing the test plot. Because I did not expect the test plot to grow, I planted another plot using the extra seed. I planted it in exactly the same way, but watered it fairly frequently. However, it was still rather late in the year. The seeds germinating at only ¼ inch below the top of the soil could hardly survive in the hot afternoon. In this area you do not have to have the germinating seed go dry for more than an hour or so before it will die. I didn't have a great deal of luck. Nevertheless, I did grow eight nice plants out of it. Some were green and some were red. We have pigweed in this area but I watched for it and kept it away from the plot.

I watered the plot quite frequently during the summer. The soil was very good as I had put manure in that same area only a few years ago. The eight plants grew to heights ranging from six feet to ten feet tall, and had many seed tassels on them. I harvested the heads before the seed was ready to shatter. However, much of the seed could be dislodged by shaking the plant. So I cut off each tassel and placed it in a large plastic bag and shook it vigorously in order to separate the seed. I then took each tassel and laid it out in a shed to dry out. This past weekend I reprocessed the tassels in order to pick up the remaining seed. The results will surprise you.

Out of the eight plants I got 4 pounds and 2 ounces of clean seed. In your reports the plants normally produce from 1 ounce to 1½ ounces per plant. Sixteen thousand plants per acre producing 1 thousand pounds works out to 1.03 ounces per plant, and 2.3 tons per acre (46 hundred pounds) comes out to 1.15 ounces per plant (64 thousand plants per acre). The yield I got was 8.25 ounces per plant which seems phenomenal, and I don't understand it. Nevertheless, it is true because I weighed it several times on an accurate scale.

I felt that the amaranth would not drop all of its seed when I first threshed it, and that is why I kept the tassels and reprocessed them. The first time I got 2¾ pounds, and the balance on the second threshing, making a total of 4 pounds, 2 ounces.

I have ground about half the seed into flour, and we have been experimenting with it. The taste of the bread is different from ordinary bread, but the texture is very good. I will keep quite a bit of the seed to plant next year. However, I am going to try to adapt this plant to our climate by treating it much the same way we would treat barley. I will plant it some time in February or March, and see that there is sufficient moisture on it most of the time until it has established itself as a young plant. Thereafter, it seems to need very little water. However, I think it is helped by water and a good rich soil. I intend to water it and fertilize it both next year. If I don't get it to germinate in February, I will try it in March; and if not then, in April.

After about four or five weeks I decided that the test plot was a complete failure. I then watered it and kept water on it fairly well for a couple of weeks. I was able to get six plants to germinate and grow even though this was in late June. These plants were all rather stunted, but had red and green tassels on them and appeared to be healthy. None of them were over four feet tall. The total production from this plot was 15½ ounces. One plant produced almost half of that.

I don't think amaranth is new to this area as I was able to find a town located in a valley quite near here (15 miles) named Amaranth that was settled around 1875.

From William O'Neill of Olympia, Washington:

We used the amaranth both as greens in salads and cooked as a side dish. It is simply great, very tasty. Even when the plants were fully mature there were always little sprigs here and there that were still tender.

I believe I recovered about 75 percent of the seeds that were produced. I also believe that a thrashing machine could be easily developed. I would almost bet that one of the old-time thrashers would do the job.

My method of thrashing — I poured the heads onto the sieve which was hardware cloth with ½-inch mesh. I used the gloves to thoroughly rub all seed and chaff off of the small branches attached to the seed head. I then sifted the grain and chaff onto a dark garbage can liner. All was placed in a light green pan. I panned the combination like "gold panning." The chaff rose to the top from which position I scooped it off and placed it in the center pan. The last part of the chaff I separated from the seed using a Shop-Vac with the hose over the exhaust vent, of course. The proper distance must be found, just far enough away to blow away the chaff but not close enough to blow out the seed. This is not difficult. In using the gloves one must really rub hard *to get the seed out of its individual little pockets. In lifting the chaff off of the top of the seed prior to using the vacuum it is easy to inspect for proper seed separation.*

From A. E. Holton, Sr. of Floral City, Florida:

I planted the sample seed on April 16. Except for some three hundred seeds picked out with a tongue-dampened needle point and planted separately for special purpose to be covered in the last part of this report. No rain fell at all for first half of this year, except for a one-inch rainfall on June 1. The ground was dry by then to two feet down. The steep, sandy hill on which my 2½ acres are located (east slope side) was insulated with dry dust which caused water to gather in sinks and low places between terraces to stand until dust was in lowest spots and then the rain went straight down as through a drain plug. This kept the seed bed from wetting deeper than the top 1½ inches of soil. This wet surface layer was bone dry again within five days. Most seeds that germinated six weeks later grew from a few inches to three feet high before they died. A few amaranth lived and a few came up at intervals in July.

Two weeks after another planting (the first week in May) the wind blew for five days from the northeast at 35 to 45 miles per hour — carrying dust, sand, and seed clouds diagonally across the hill and countryside. I imagine some seed were buried too deeply in low places between terraces (six feet apart to prevent erosion) to allow germination, but those that stuck on terrace tops germinated and four hundred to five hundred plants survived for a while and about half of them survived to such maturity as they were capable of.

Eighty-four plants grew to seven- to ten-foot heights and produced heads heavy enough to be worthy of seed harvesting for propagation purposes. Eight pounds and 7 ounces of seeds were harvested from these plants. According to my figures, that amounts to 1.6 ounces per plant.

Several large, dark red seed heads were saved for decorative seed propagation and most of the lesser-developed amaranth were used for chicken and human food experiments. Chickens liked the leaves as well as any high protein vegetation and better than most. They ate the leaves even when they were large, old, and bitter as quinine. The tiny grain of the matured heads were sought after as if they were micro-manna from heaven, chicken heaven, that is. As human food, the small leaves of early development were as you reported: they tasted like mustard greens and were delicious. The slightly larger leaves lost this flavor and tasted like some kind of hay and as they became larger still, they turned more and more bitter. The first stage of bitterness made them taste like old-time cowhorn turnips but later they became too bitter for my taste buds to tolerate.

As for the amaranth grain as human food: I know now what God fed the children of Israel in the wilderness after they left Egypt — it was snow storms of popped amaranth grain. The sun must have been hot enough to pop the windblown grain after it hit the ground. The grain was as good as you said it was and in all the ways you said it would be, but the limited amounts I had to experiment with did not allow me to go as far in experimenting as I would have liked to go. The experiment I liked best was when I used a cup half-full of half-popped, half-toasted grain, mixed with an equal amount of pancake batter for griddle cakes or, as we call them, flapjacks. This makes flapjacks nonflaps and greater jacks — you ought to try it.

. . . I planted the amaranth seeds on April 16 and had fair success with them which furnished the bulk of the eight pounds and seven ounces of seed for next year. Five different types of plants were grown from these seeds — most produced white, giant amaranth averaging four to ten feet tall; a few were bright red in color, some were pink or had part red and part white heads. A number of the plants were either throwbacks to wild species with wide-spreading branches and very small seed strings, or they were hybrids of giant wild species.

There was one special hybrid plant that grew and matured much faster than any of the others. It had a pale mother-of-pearl pinkish color on its 20-inch-long and 14-inch-diameter head, which was compact and very heavy with immature seeds when two months and ten days old. This plant also had 19 large, wide-spreading branches which bore heads as large as many of the other plants that matured later. On June 26 when I destroyed this plant accidentally, while working a corn row too closely planted to it, I collected the seed heads and put them in the hothouse to dry, but five days later, on July 1, when I started rolling, rubbing, and beating the seeds out it was evident that they were not all mature seeds, mostly in the "whole corn" stage. Some were drying out into tiny, flat, tan-colored discs with a brown spot in the center.

There were 11½ ounces of these green seeds, and while I knew that many were too immature to germinate, I was positive that some would — so I planned immediately to use these seeds in a program of selective breeding to get a special high-producing, fast-growing variety of giant amaranth.

I planted these seeds on July 1, between the dying and stunted corn hills by dropping thumb-and-forefinger pinches of seeds on the brown spots where holes had been dug and shovels of chicken manure had been lightly covered earlier in the year and the single rain of the year had caused the manure stain to show, but it had come too late to save the corn. On July 2, 3, and 4, it rained intermittently each day and the summer wet season was on, such as it was. Our rainy season this year was lighter than usual but lasted through September, which was unusual. The weather bureau reported that this was the second worst drought year in weather reporting history for this area, being 2½ inches short of the year's annual rainfall.

The very special amaranth seed came up immediately in wads of 25 to 35 plants to the spot. The green manure crop of vetch, beggar weed, and other plants was already several inches high and foretold the coming battle of competition for sunlight and food that was to be the lot of these amaranth plants.

My plan called for a law-of-the-jungle, survival-of-the-fittest program, that would allow only the fastest-growing, most resistant to insect/pest type of seed plant to reach maturity. A six-foot-deep mass of plant growth was what these plants had to conquer before they could reach sunlight. The 91 largest heads, even though the leaves were stripped by insect pests to skeletons, still produced 0.64 of an ounce of seed per head. Total seeds produced from these second-crop plants amounted to three pounds and ten ounces.

Insects and worms observed were corn ear worms, corn borers, stink bugs, tiny brown, tan, and black weevils; a small, flat ½-inch-long web-worm and many others not previously noted in the area, completely, or partially destroyed a high percentage of the plants and seed heads. A great many of the amaranth plants from this July 1 planting, reached a height of only 20 inches to three feet and produced nothing, as was to be expected. The three pounds and ten ounces of special seeds will be planted next April 15, to continue the experiment for development of a more adapted, high-producing, and fast-growing variety of amaranth for this area.

From Jim and Linda Zalusky of Dayton, Ohio:

Our overall impression was that amaranth was easy to grow, withstood dry weather well, and did not seem to be bothered by insects. We have begun to use the amaranth grain and are pleased with its taste and texture.

We felt that two factors need the most attention in developing amaranth for the gardener:

1. *Breeding of amaranth for uniformity and higher yields.*
   We received seed from another supplier which we planted under similar conditions. Its size and yield were definitely inferior. Also, even within the Rodale experimental plot, great variation in size, color, flowering time, and yield was observed.
2. *A better method of threshing.*
   As we threshed the grain, it seemed to us that a good opportunity existed to adapt Rodale's pedal power idea to amaranth grain threshing. For those of us who would like to raise 30 to 50 pounds of grain a year, such a unit in kit form might prove valuable.

PS We would like to take part in any further amaranth experiments.

From John Pearl of Bowier, Maryland:

Approximately 90 percent of the amaranth seed germinated (planting made in mid-May). Plants were very vigorous (shortest plant was 6′0″; tallest plant was 8′11″), and relatively unaffected by pests. Two varieties of plant were evident: one with yellow/maroon blossom heads, the other with green heads. The first variety bloomed early in July and tended not to grow side shoots unless plant was broken; the plants were relatively brittle and tended to snap off about 40″ from the ground in rough weather. The second variety bloomed later (mid-August) and tended to grow several strong lateral stems from near the base of the plant; these plants were much more tolerant of rough weather and ultimately tended to be pushed over rather than broken.

I had intended to leave all of the plants unstaked to determine their ability to withstand wind. However, after receiving your letter of July 1, I did tie up some of the plants about 3 feet from the ground. In subsequent storms, both the staked and unstaked yellow/maroon plants broke off very consistently about 3½ feet from the ground.

The amaranth leaves tasted quite good. We particularly enjoyed them boiled as greens, and in the green rice casserole described in your Amaranth Round-Up *booklet.*

I had some difficulty determining when seeds were ripe. The green variety blossomed over a month later than the yellow/maroon variety, but by the end of September the seed heads seemed to be shattering on both. Ultimately only two green-headed and one yellow/maroon-headed plants were harvested for measurement; bitterness in the seeds collected from the green-headed *plants suggests that they were not fully ripe. The remaining plants were knocked down and broken off (surprisingly uniformly at about $3\frac{1}{2}$ feet from the ground) by bad weather.*

Although only three plants were used for the documented harvest, heads of many of the broken plants were also collected, since the broken tops usually remained alive; in some cases the plants were merely bent sharply down at a 120° angle, but did not snap. Unfortunately, the "extra" harvest was lost to mildew when I unsuccessfully tried to dry it in the basement; though the seed heads were shattering, they were still much more succulent than I suspected.

The drying of the plants can be materially speeded up by splitting the stalks twice lengthwise; the outer husk of the stem is so woody that it is virtually impermeable to moisture.

There were very few pests. Although quite a few holes up to $\frac{3}{8}$ inches in diameter appeared in the leaves, plant growth and health seemed unaffected. Deer were a serious pest in the remainder of the garden, but did not bother the amaranth; breakage of plants was almost certainly due to heavy wind and rain, as evidenced by extensive damage following a storm. Insects revealed by occasional casual searches were:

June 19: *Japanese beetle* (Popillia japonica), *striped cucumber beetle* (Acalymma vittata), *and ladybug* (Hippodamia convergens; Adalia bipunctata)

July 9: *stalk borer* (Papaipema nebris), *green June beetle* (Cotinis nitida), *Japanese beetle, ladybug*

October 1: *spotted cucumber beetle* (Diabrotica undecimpunctata howardi), *squash bug* (Anasa tristis), *stink bug* (Euschistus survus), *tarnished plant bug* (Lygus lineolaris), *aphids (species unknown), stalk borer, striped cucumber beetle*

Only the stalk borer was obviously damaging the plants; tunnels up to 18 inches long were found in the stalks of three. Due to the large size of the plants, the borers seemed to have no deleterious effect, however.

From Gam B. Louie of San Carlos, California:

I have been raising amaranth in my garden for the last five years. When it is about 18 to 24 inches tall, I cut off the top, leaving four or five leaves behind for future branches. It takes only a week for it to bush out and produce more cuttings for the table.

I have given this vegetable to many people who all tell me it tastes better than spinach, and some say the red ones are sweeter than the all-green ones.

One of my neighbors, who is from Greece, tells me that in Greece these vegetables are served in the finest hotels and considered a delicacy in that country. She usually cooks it with lamb, but our people cook it with a bit of meat in soup after it is tender, add seasonings to taste and break an egg and stir it into the soup. Or, brown a clove of garlic in oil for flavor, add the amaranth, stir, add seasonings to taste and cook until tender.

I do hope your readers try cooking and enjoying the leaves and stems of the amaranth vegetables. We Chinese call it "hun-toy."

From Henry T. Brown of Webster, New York:

Thanks to your contribution of seed and instructions, I have raised amaranth in Jamaica. There are now a dozen Peace Corps Volunteers and Jamaican agricultural people planting seed I produced there.

As a Peace Corps Volunteer, I found the amaranth

project one of the most productive of any of our ventures. Not only did it excite community interest, but I was beseiged with questions. That's something the people usually shy from.

From Lee Price of Woodland, California:

Encouraged by the characteristics of grain amaranth, I purchased seeds (78¢ for ⅛ ounce) and planted them. Even without the benefits of irrigation, rainfall, or fertilizer, my amaranth plants produced 2.35 ounces per plant.

I have made cookies, bread, hot cereal, "corn" bread, and other foods from the seeds. To me, the best thing to be made from them is a flour which, in combination with some other ingredients, makes an instant "milk" for feeding the hungry children (and adults) of the world. I also made tortillas from amaranth.

My "field" (nine hundred plants) of amaranth produced 128 pounds of seeds. Seed companies now offer a thousand seeds for 65¢. At that rate, my crop has a value of $7,987.20. At a production of a ton per acre, a field of one hundred acres would have an astronomical value! Of course, all those figures are fun-calculations but it does point out that amaranth seeds are valuable.

In my quest to grow large crops of amaranth and get its high protein into the hands of the under-nourished peoples of the world, can you offer any suggestions of how to go about it?

With that last question, Lee Price condensed the entire five-year effort at Rodale and other organizations and agencies into one sentence. It's a matter of getting the new crop from "there to here," and, as you can tell from reading these few sample letters, it is a process which has been helped immeasurably by the thousands of Reader-Researchers who have responded, and will continue to respond, to the call for still more tests and still more information. That's one of the ways "to go about it."

One of the many tangible results of the Reader-Research project has been the development of a basic set of instructions for planting, cultivating, and harvesting amaranth. For any of you who would like to try, it is in Appendix A. Remember, thousands of backyard gardeners helped to write it.

With a packet of amaranth seed and a bit of earth anywhere in the nation, you can span millennia and help revive a process and a crop that was part of this earth near the port city of Veracruz thousands of years ago. Or, if you would like to volunteer, you can enlist in the Rodale Reader-Research corps. If you do, you'll be sent some of the new varieties of amaranth seed that Skip Kauffman, Laurie Feine, Dick Harwood, and all the other people at Rodale have helped bring into being over the years. Because one thing is certain: work on the amaranth — that versatile, nutritious, historic, and symbolic plant of five continents and many centuries — has only just begun.

It will get from "there to here," it will help feed the world's hungry, it will be restored to the place it once commanded in the nutritional and horticultural scheme of things. Having read its story, can you doubt its eventual success?

Appendix A
Growing Guidelines

Grain Amaranth

Grain amaranth can be either transplanted or direct-sown. Choose either method for a successful crop.

Transplanting:

For best results, start your amaranth plants no more than a month before your last frost date or just before sweet corn goes in the ground. Plan to use peat pots, or flats which will allow a good amount of soil around roots when transplanting.

Mix some of the seeds with potting soil (½ cup). Fill peat pots or whatever you prefer to use with moist soil. Plant a pinch of the seed/soil mixture (at least 3 seeds to a hole) in a small depression. Cover with about ⅛ inch of moist potting soil. Sprinkle lightly with water every few days to keep the soil moist.

Thinning should be done progressively as the seedlings grow. The most vigorous plant is left in each hole while others are pinched off. When thinning, do not pull up the plants by the roots, as this disturbs the root systems of the plants that remain.

When the last frost of the spring is past and the ground is warm, rototill your garden plot at least once and transplant your amaranth seedlings. There have been a number of successful spacings used in growing grain amaranth. At the Research Center in 1978 good yields were obtained when plants were 40 inches apart between rows and 20 inches within rows. In 1979 spacings of 30 inches × 30 inches will be used for ease of cultivation.

Direct-Sowing:

When the ground is thoroughly warm (about 2 weeks after sweet corn is planted in your area), it's time to direct-sow your amaranth. Prepare the plot(s) so that a fine tilth is obtained.

After preparing the plot, make a guide string to help keep your planting rows and holes straight and uniformly apart. See transplanting directions above for spacing guidelines. With a stake tied to either end, stretch the string across one end of your plot. Smooth the soil along the length of the string with a flat board. With a finger, make small depressions beside the guide string on the smoothed soil. Put a pinch of seed/soil mixture (at least 3 seeds in a pinch) in each depression and cover very lightly with dark, moist potting soil to mark the holes. Then move the guide string with the pegs to the next row and repeat this procedure until the plot is planted. Sprinkle the plot lightly with water every few days to keep it moist.

Follow the instructions for thinning that appear under Transplanting.

Culture:

You can mulch your amaranth plots about a week after thinning (or immediately after transplanting). Use composted material, hay, grass clippings, wood chips, newspapers, or whatever you normally use for mulch. Do it uniformly to a depth of about 6 inches on all plots. This will help retain soil moisture and a more even soil temperature.

Water your plots as often as you think is necessary. Amaranth needs water for germination and during its seedling stage; after that, it will flourish even in droughtlike conditions.

Approximately 60 to 70 days after planting, you can expect to see your plants begin to flower. This will be evident by a small inflorescence — yellow, magenta, or a variegated combination of both — formed at the top central part of the plant.

Harvesting:

Seed heads should be harvested at the early stages of maturity. Pick a small portion of seed head and rub it lightly between your fingers. If the seeds dislodge easily, it could be time to harvest. To test the seed, try the "dough stage" test. Chew on a seed. If it's soft like dough, it's not quite ready to be harvested. When the seed is firm, then it is mature and ready to be harvested. Plants should not be allowed to become too dry in the field because too much seed will be lost due to shattering.

Begin harvesting after morning dew has evaporated from plants. Cut off the seed heads (those on lateral branches as well) and spread them out on a big sheet in the sun (black plastic works well) or a dry section of your greenhouse, garage, barn, basement, or attic to dry. When they're fairly dry, strip off the leaves and store the heads, wrapped in bundles of three heads, in good burlap bags until thoroughly dry, and you're ready to thresh and clean the seed.

Remove the stalks from the field and compost them. Shredding the heavy stalks will accelerate decomposition.

You can begin processing the seed when the plants are thoroughly dried. First, beat the bags with a broom or rug beater to loosen the seeds and parts. Then, wearing stout gloves, rub the mixture through a ⅛-inch screen; this will separate the seed and chaff from the bigger pieces. Sieve this finer part through 16-mesh window screen, grain sieves (Burrows Grain Dockage Sieves — 1/16 or 1/18 — work well), or a flour sifter. The end product will be fairly clean.

Some of the Reader-Researchers who grew amaranth were satisfied with gently blowing away the chaff from the grain. However, others found that rigging up a mechanical blower like a fan, hair dryer, or vacuum cleaner (with hose over the exhaust vent) worked well to clean the chaff from seed. The challenge is to calculate the correct distance and air speed.

Before milling the grain for flour, you can rinse it free of dirt and lightly toast it on a cookie sheet in your oven. It will grind into a fine flour using a small coffee grinder, blender, or grain mill. Another good way to use the seed in recipes, especially if you don't have a mill, is popped. See Appendix B for recipes.

Vegetable Amaranth

Vegetable amaranth is a warm weather crop. It likes warm, moist soil and full sun for best growth. Plan the planting to follow your spinach crop or to begin after the squashes begin growing rapidly — when the soil and air are warm.

You can either transplant or direct-sow. Either method should produce a successful crop.

Transplanting:

Sow the seed thinly in flats 2 to 3 weeks before the outdoor planting date. Transplant the seedlings when they are 2 to 3 inches tall into a bed or relatively fine tilth. Set them approximately 3 inches apart. Thin to 6 inches apart when 6 inches tall and again if crowding occurs. Do not pull up the plants by the roots as this will disturb the root systems of the plants that remain. Pinch them off instead. The smaller plants, as thinned, can be cooked and eaten whole.

Direct-Sowing:

Sow seed thinly in rows 12 inches apart or broadcast (mix with sand for easier broadcasting) in beds of good tilth. Cover with ⅛ to ¼ inch of light soil or potting mix. You can expect the seedlings to take from 4 to 14 days to emerge, depending on soil temperatures. Keep the soil moist but not soggy.

Thin seedlings to 3 inches apart when 3 inches tall, 6

inches apart when 6 inches tall, and again later, if crowding occurs.

Culture:

Water the plots as often as you think necessary. The vegetable varieties seem to do best when kept well watered.

Some varieties may begin to flower in 35 to 40 days; others probably will not flower throughout the season. Remove the flower heads as they appear, as this will prolong the production of edible leaves.

The first crop of thinnings should be ready 3 to 4 weeks after seedlings emerge. If the plants are adequately spaced to avoid crowding, pluck the tender tips of the central stalks or branches and let the plants grow new branches before harvesting again in 2 or 3 weeks.

Harvesting:

When you're ready to clear out garden space, you can pull the whole plant up or cut it at ground level. This "whole plant" harvest is similar to harvesting spinach or leaf lettuce.

Grain Cleaner*

Cleaning amaranth grain has always posed a problem for the staff at the Organic Gardening and Farming Research Center and probably for other people for centuries. Blowing off the debris from the grain with your mouth works fine, but it has obvious problems, and buying a commercial grain cleaner could run into a small fortune.

We came up with this grain cleaner, which does a good job on amaranth grain and also many other types of grain. To use the cleaner one drops the uncleaned material in front of the air flow which is strong enough to blow away the debris but not the grain.

---

*Written by Barney Volak.

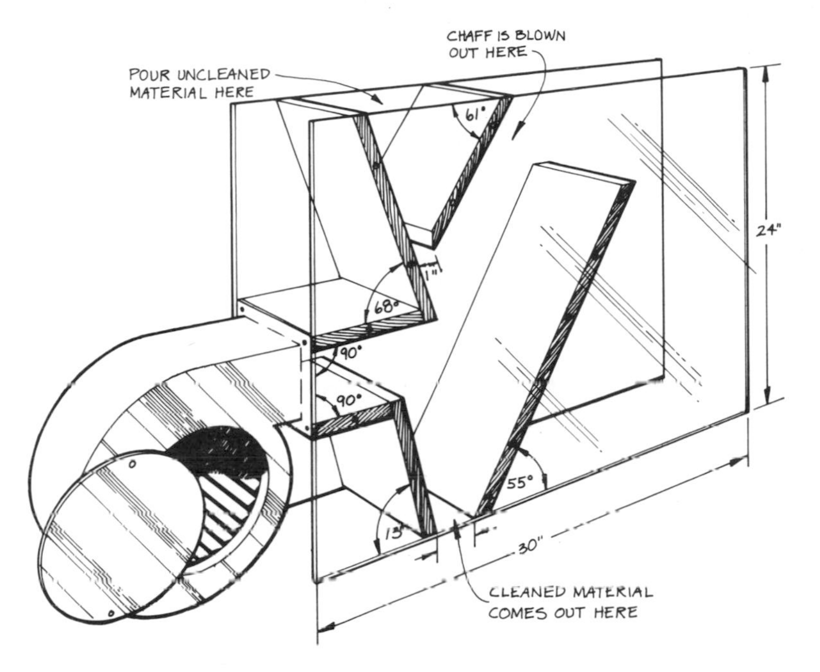

Materials

1 4′ × 30″ × 3/8″ plywood
1 6′ × 6½″* × ¾″ pine board
1 blower
26 #8 wood screws, 1¼″ long
glue

Construction

1. The first step is to cut the 4′ × 30″ × 3/8″ plywood in half to get 2 2′ × 30″ pieces.
2. Next, determine the size of the blower's air outlet. We used a Dayton Shaded Pole Blower, 1/10 h.p.,

*The width of the board is determined by the width of the blower outlet.

1,570 rpm, with free air movement at 525 CFMs. (This blower has enough air movement to clean larger grains also.) The blower air outlet will determine the width of the pine board used in the next step. (Example: Our air outlet was 6½″ wide, so we made the board 6½″ wide.)

3. After you have the right width board, cut it to the proper lengths using diagram below.

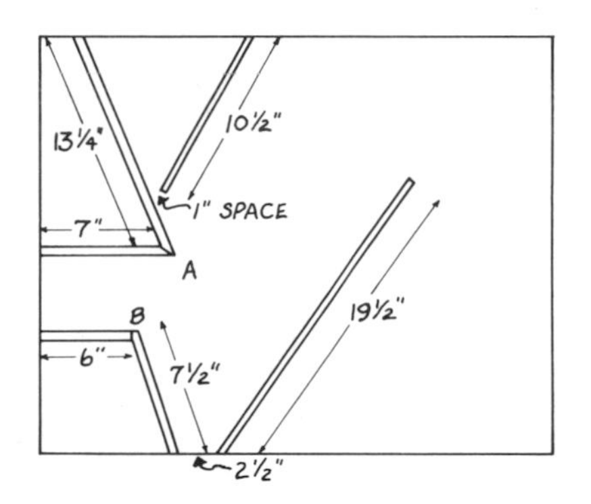

4. Arrange the pieces you have cut on 1 piece of plywood, using the diagram. To receive proper air flow, surfaces A and B must be smooth, so there is a direct flow of air with no disturbances. To fit these surfaces together properly, use a rasp bar and plane to the proper angles to insure the best fit.
5. Next, glue and screw the pieces to the plywood. When glue has dried, attach other piece of plywood.
6. Attach the blower to the framework.
7. To clean different size grain you will need an air inlet adjustment, which is simply a tin plate or piece of cardboard over the air inlet. This controls the amount of air flow on the falling material. It will take a few times to find the proper opening.
8. We set the cleaner directly over a hole cut in a table

and placed a bucket under it to catch the seed. You can also place the cleaner on blocks and catch the seed in a pan.

When cleaning your grain, remember to pour it slowly. You may have to run it through the cleaner 2 or 3 times if it has a lot of debris.

Appendix B
Amaranth in the Kitchen*

In this appendix we come to the most important reason for growing your amaranth — to eat it. Both the grains (the seeds) and the greens are delicious and can be prepared in a number of ways.

The Grains

To prepare the grain, you must first be sure that it is very clean and free from any chaff or odd specks of dirt. Take the extra time to shake it in a very fine strainer. This will minimize the grittiness in your amaranth dishes.

Over one hundred years ago, Sylvester Graham, who was the originator of breakfast cereals and gave his name to graham flour, exhorted housewives to wash and dry wheat to obtain a sweeter loaf of bread. You can do this, too, with amaranth grain by washing it very thoroughly in a fine strainer under running water. Dry it on baking sheets at room temperature, then store it in moisture-proof containers until needed.

Due to the grain's relatively high oil content, when milled alone in large amounts it will clog a stone grain mill, such as Magic Mill. However, grinding a small quantity, such as a cup of flour as you need it, works fine in any mill, including the

*Written by Susan Asanovic.

stone-grinding models. Also, since the seeds are so tiny, many will slip through unground. You may lightly toast the grain first, then grind it after thorough cooling; or grind the untoasted grain along with an equal amount of wheat, millet, corn, or whatever other grain you plan to use with it. For small amounts (less than half a cup), an electric seed mill or coffee grinder is very practical, and when using either one, no pre-toasting or combining is necessary. However, toasting the grain after grinding gives the amaranth a nutty flavor.

The cooks in Rodale's Experimental Kitchen have found that popping amaranth grain before grinding it eliminates any problems with grittiness or clogging that might be present when grinding the grain raw. Directions for popping the grain are presented near the end of the grain recipes that follow this section. Popped and ground amaranth grain can be used in place of any raw or toasted amaranth flour in any recipe.

In a bread, cake, or muffin recipe, or any other recipe for a food that must rise, you can substitute ½ cup of amaranth flour for ½ cup of the other flour for each 4 cups of any whole grain flour. For example, if a recipe calls for 8 cups whole wheat flour, you may use 1 cup of amaranth flour and 7 cups of wheat flour in all. In any recipe for a food that need not rise, like crepes, pancakes, or cookies, amaranth grain *that has been popped before grinding* can be substituted for all or any part of the wheat or other flour required.

Amaranth is unique among grains in that it boasts a "protein score" higher than that of cow's milk, which means that your body can use amaranth protein more efficiently than milk. However, it is weak in one essential amino acid, tryptophan. Always try to combine amaranth grain with food high in tryptophan, such as beans, nuts, seeds, wheat, cheese, eggs, or brewer's yeast. It is strong in one essential amino acid, lysine, which is characteristically low in grain such as wheat. A small amount of amaranth will complement these grains and other plant proteins very well.

APPENDIX B

The Greens

The greens need only a brief washing before using them in salads or cooking. Use them fresh from the garden for best flavor and vitamin value; only pick what you need for a meal. Greens are generally tasty up to approximately 4 inches long, after which they become bitter and tough. You can harvest them throughout the season as new leaves are always growing along the stalk. When they are very young, you may use them as a green in mixed salads. If you find yourself with too many at one time, you can blanch them for 2 minutes, cool quickly, and then freeze them in meal-size portions. In the recipes that follow, if you are using frozen amaranth use half the amount called for fresh.

Amaranth greens are outstanding among vegetables in that they supply vitamin A, calcium, iron, and other minerals. Cook them just like young leaf spinach, adding ¼ to ½ cup of water to a pot, bring the water to a boil, add amaranth, cover, and steam for 5 minutes. Cook in as little water as possible in the shortest time to preserve the vitamins. As the leaves mature, you will find that any milk-base or cheese sauce will tone down a strong flavor. Some find that tamari soy sauce or miso or even vinegar will have a similar effect on the mature plant acids. You can use amaranth leaves instead of spinach or chard in your favorite recipe. You may also want to try the recipes that follow, some created just for amaranth and all tested in Rodale's Experimental Kitchen.

Grain Recipes

Grains and Breakfast Foods

Amaranth Mixed Grain Pilaf

This is a delicious and simple way to prepare the grain.

- 2 tablespoons amaranth grain
- 1 cup less 2 tablespoons millet
- 1 tablespoon sesame oil
- 1½ cups water

Toast amaranth briefly in heavy-bottom saucepan. Do not let it begin to pop. Add water and oil and bring to a boil. Slowly pour in millet, cover, and reduce heat to minimum. Steam 30 minutes. It is important to stir the pilaf with a chopstick or wooden fork and fluff it well before serving, as the heavier amaranth tends to sink to the bottom.

This is delicious as a hot main dish. You may also serve it cold as tabouli, mixed with a parsley vinaigrette and garnished with mint leaves, tomatoes, and black olives. You can also serve room-temperature pilaf with yogurt and snipped fresh herbs. Just add enough yogurt to moisten and lots of herbs, such as parsley, chives, mint, basil, or tarragon.

3 servings

Amaranth Yogurt Pancakes

- 1 cup yogurt
- ¼ cup water
- 1 teaspoon baking soda
- 1 egg
- 1 tablespoon vegetable oil
- ½ cup unbleached flour
- ¼ cup whole wheat flour
- ¼ cup amaranth flour
- Fresh Strawberry Pancake Topping (see recipe below)

Combine yogurt, water, and baking soda. Beat together egg and oil, and add to yogurt mixture. Add flours and stir in lightly. Bake on hot, oiled griddle. Serve with Fresh Strawberry Pancake Topping.

12 3-inch pancakes

Fresh Strawberry Pancake Topping

1½ tablespoons cornstarch
3 tablespoons honey
2½ cups strawberries, washed and hulled

Mix cornstarch and honey in saucepan. Mash strawberries, add to cornstarch mixture, and cook over low heat until clarified and creamy thick (approximately 5 minutes).

1½ cups

Hot Breakfast Amaranth Cereal

Prepare your favorite hot cereal: oatmeal, cornmeal mush, cracked wheat, or other. Five minutes before serving, stir in 1 tablespoon of amaranth flour for each serving. Simmer very gently for a few minutes. It is a great taste and nutrition booster for any cereal.

Mixed Grain Pancakes

½ cup whole wheat flour
¼ cup cornmeal
¼ cup amaranth flour
1 egg
1 cup buttermilk
2 tablespoons vegetable oil
1 tablespoon honey
1 teaspoon baking powder
½ teaspoon baking soda

Combine ingredients in a large bowl, and mix until smooth. Bake on a hot, oiled griddle.

8 to 10 3-inch pancakes

Oatmeal and Amaranth Cereal

4 cups water
1 cup oatmeal, regular
½ cup amaranth flour
½ cup raisins
1 tablespoon honey
1 tablespoon sorghum molasses
2 tablespoons wheat germ, untoasted
2 tablespoons sesame seeds, untoasted
¼ teaspoon cinnamon
milk for serving

Bring water to boil in a 2-quart saucepan.

Stir oatmeal into boiling water. Add amaranth flour and raisins. Cook for 10 to 15 minutes, until the oatmeal is tender.

Add honey, molasses, wheat germ, sesame seeds, and cinnamon. Serve warm, with milk.

4 servings

Whole Grain Easy Amaranth Cereal

- 3 tablespoons amaranth seeds
- ⅓ cup boiling water
- pinch of kelp

Simply pour the seeds very slowly into boiling water, season with kelp, cover, reduce heat to minimum, and steam at least 20 to 30 minutes. Unlike other whole grains, amaranth stays chewy, even when well done. Add water if necessary as water is absorbed.

1 serving

Breads

Amaranth Bread

- 3 cups whole wheat flour
- 2 envelopes dry yeast
- ¼ cup carob powder
- ¼ cup wheat germ
- ¼ cup brewer's yeast
- 2 cups water
- ⅓ cup unsulfured molasses
- 2 tablespoons butter
- 1½ teaspoons honey
- 2 cups ground amaranth flour

In large mixing bowl, combine 2 cups flour, yeast, carob powder, wheat germ, and brewer's yeast. In saucepan combine water, molasses, butter, and honey; heat on stove until butter melts. Let cool slightly and gradually add to dry mixture, stirring by hand.

Slowly add the amaranth and remaining whole wheat flour to make a soft dough. Turn onto a floured board and knead until soft. Cover dough ball with mixing bowl and let rest for 20 minutes. Punch down and divide dough in half.

Roll out, fold, and shape into 2 loaves, and place in greased loaf pans. Place in unlit oven with a pan of hot water and let rise until double (about 1 hour). Remove water pan and bake at 400°F. for 30 minutes. Remove from pans to wire racks to cool, covering with clean cloth for tender crust.

2 loaves

Amaranth Brown Bread

- ⅔ cup butter
- 1 cup molasses
- 1 cup sour milk, buttermilk, or yogurt
- 1 cup chopped raisins or currants
- ½ cup chopped walnuts
- 2 cups whole wheat flour
- ½ cup amaranth flour
- 2 teaspoons wheat germ
- 1 teaspoon baking soda
- ¼ teaspoon nutmeg
- cream cheese for serving

Cream the butter with molasses. Add the sour milk, buttermilk, or yogurt and chopped raisins or currants and chopped walnuts. Combine the flours, wheat germ, baking soda, and nutmeg. Stir into molasses and sour milk mixture.

Place in 5 oiled soup or other small cans. Cover tightly with oiled brown paper or foil. Set in a large pot and add boiling water until it comes halfway up the cans. Cover the pot with a close-fitting lid. Bring to a boil and steam for 1 hour. Unmold, slice, and serve warm with cream cheese.

5 small loaves

Amaranth Corn Muffins

- ⅔ cup whole wheat flour
- ⅓ cup soy flour
- ⅔ cup corn flour
- ⅓ cup amaranth flour
- 1 tablespoon baking powder
- ¼ cup honey
- 1 cup coarsely chopped cranberries, blueberries, or raspberries
- 1 cup milk
- ⅓ cup melted butter
- 2 eggs

Grease 12 muffin cups. Combine flours with baking powder in a large bowl. Add honey to berries and add to flour mixture.

Measure milk in 2-cup measure. Add melted butter and eggs; beat with fork to mix well. Make well in center of flour mixture. Pour in milk mixture all at once; stir quickly with fork just until dry ingredients are moistened. Batter should be lumpy. Quickly fill muffin cups with batter, about two-thirds full.

Bake 25 minutes at 400°F. or until golden brown.

12 muffins

Amaranth Crepes

These are loaded with high-quality protein and can certainly be served as the main dish of any meal.

- 2 cups milk, cow or soy
- 4 large eggs
- 2 tablespoons vegetable oil
- ½ cup amaranth flour
- 1⅓ cups whole wheat pastry flour
- 1 teaspoon organic orange or tangerine peel, grated (optional)

Put milk, eggs, and oil in the blender, then add flours and grated peel (if desired). Blend until smooth and refrigerate for at least 2 hours. Fry in a hot, lightly oiled 7-inch skillet, using about ¼ cup of batter for each crepe. Add more oil as needed between every second or third crepe.

For a variation, add 2 tablespoons of honey to the batter and roll up with a fruit compote for breakfast, or wrap around raw or lightly cooked vegetables for lunch or supper.

about 16 crepes

Amaranth Tostados

A fast snack bread to accompany curries or chilies.

- 1 cup amaranth flour
- 1 cup finely ground cornmeal
- 2 tablespoons corn oil
- ⅔ cup boiling water

Mix dry ingredients. Put oil in another mixing bowl, add boiling water, and slowly pour into dry ingredients, stopping when dough just hangs together. Gather into a ball and when cool, cut into egg-size pieces. Roll out into 6 circles, about 7 inches each, quite thin, on a floured board or cloth.

Bake on a medium-hot, cast-iron skillet or griddle until dry and slightly brown; turn and repeat. Stack and cover with a tea towel until ready to serve. These can be reheated by sprinkling with a little water and warmly briefly in foil in the oven.

6 tostados

Fluffy Dumplings for Soup or Stew

- 1 pint of seasoned stock (chicken or beef)
- ½ cup cornmeal
- 1 tablespoon amaranth flour
- 1 whole egg

Bring stock to a boil and slowly stir in the cornmeal and amaranth flour. Remove from heat source and beat vigorously until all of the mixture is blended smoothly and leaves the sides of the pot. Cool for 5 minutes. Add egg and beat well until it is blended into the mixture. Cool 5 minutes more. Working with 2 spoons, drop dumplings about the size of walnuts onto a plate or cookie sheet.

Bring soup or stew to the simmering point in a broad pot with a tight lid. Remove the lid, and carefully slip the dumplings into the hot liquid, taking care to place them next to one another but not on top of one another, as quickly as possible. Close the lid tightly, and allow the dumplings to cook for 5 minutes.

15 walnut-size dumplings

No-Knead Whole Wheat Bread

7 cups whole wheat flour
1 cup amaranth flour
2 envelopes dry yeast
1 cup lukewarm water
1 tablespoon honey
4 tablespoons molasses
1 cup warm water
2 cups warm water (approximately)

Place whole wheat flour and amaranth flour in large bowl and set it in a very low oven for about 20 minutes, to warm flour and bowl. If it is a gas oven, the pilot light will give sufficient heat; if electric, set at lowest temperature.

Dissolve yeast in 1 cup lukewarm water and add honey. Mix molasses with 1 cup warm water. Combine yeast mixture with molasses mixture and add to warmed flour. Add enough warm water to make a sticky dough (about 2 cups).

Oil 2 large loaf pans, at least 9 × 5 × 3 inches or larger, or 3 small loaf pans, and put entire mixture directly into pans. No kneading is involved. Let rise 1 hour.

Preheat oven to 400°F. Bake for 30 to 40 minutes or until crust is brown. Remove pans from oven and leave to cool on racks for 10 minutes. Remove loaves from pans and continue to cool on racks before slicing.

2 large or 3 small loaves

Orange-Molasses Muffins

1 cup whole wheat flour
3/4 cup amaranth flour
2 teaspoons baking powder
1 egg
1 tablespoon orange peel, grated
1/2 cup orange juice
1/2 cup milk
2 tablespoons vegetable oil
2 tablespoons molasses
1/2 cup raisins

Grease 12 muffin cups.

Combine dry ingredients. Beat egg and add orange rind, orange juice, milk, oil, molasses, and raisins. Combine dry and wet mixtures. Drop into muffin cups. Bake at 400°F. for 20 minutes.

12 muffins

Prune Nut Bread

- 1 cup whole wheat pastry flour
- ½ cup amaranth flour
- ⅓ cup skim milk powder
- 2 teaspoons baking powder
- 1 cup sunflower seeds
- 1½ cups pitted, stewed prunes and juice (see recipe below)
- 3 tablespoons vegetable oil
- ⅓ cup honey
- ½ cup raw wheat germ

Sift first 4 ingredients into a large mixing bowl. Add sunflower seeds and mix until they are coated with flour. Add prunes, oil, honey, and wheat germ to the mixture. Stir with no more than 40 strokes. Line the bottom of a 4 × 9-inch loaf pan with waxed paper and grease well. Pour the batter into the pan, forcing it into the corners. Make an indentation lengthwise through the center. Bake at 350°F. for approximately 1 hour.

1 loaf

Stewed Prunes

Place one 12-ounce box of dried prunes in a small saucepan. Add enough water to cover them, cover, and cook over medium heat for 5 minutes.

Tortillas

> He also sells large thin tortillas — tortillas cooked in coals; . . . tortillas made of amaranth seeds and of ground squash seeds and of green corn and of prickly pears. . . . Fray Bernardino de Sahagun, quoted by Diana Kennedy in *The Tortilla Book* (Harper and Row, 1975).

- 1 cup cornmeal
- 1 cup cold water
- 7/8 cup amaranth flour, toasted gently
- 1 tablespoon olive oil

Add cornmeal to cold water, and slowly bring to a boil. Constantly stirring to prevent sticking, cook over very low heat until thick and a mass begins to form. Cool, then add oil and enough toasted amaranth flour to make a stiff dough.

Pinch off pieces the size of a golf ball and roll quite thin between two pieces of plastic wrap, or flatten in a plastic-lined tortilla press. Place a tortilla on your palm, leaving a piece of plastic on the open side, then peel off the second sheet of plastic and throw your tortilla onto a heated, dry iron griddle or skillet. Bake until dry and beginning to brown; turn and bake until dry, but still pliable. Stack and cover to keep soft.

Too dry a dough will cause tortillas to crack when rolled out; too much moisture will make them stick to the plastic. Adjust liquid or dry ingredients accordingly. Experience will dictate the correct amount.

12 tortillas

Whole Wheat and Amaranth Rolls

- 1 cup milk
- 3/4 cup butter
- 2 tablespoons honey
- 1 envelope dry yeast
- 1/4 cup warm water
- 3 1/2 cups whole wheat flour
- 2 eggs, room temperature
- 1 cup amaranth seed, popped
- 1 1/2 tablespoons butter, cut into 1/4-inch cubes

Heat milk, butter, and honey until butter melts; cool to lukewarm. In a large bowl, dissolve yeast in 1/4 cup 110°F. water. Combine yeast and butter mixtures. To this, add 2 cups of the whole wheat flour and beat on the slow speed of an electric mixer for 3 minutes or until smooth. Beat in the eggs. Add amaranth seed and slowly add 1 cup of the remaining flour, beating slowly another 2 minutes. Add remaining ½ cup flour and stir well with a wooden spoon.

Place the mixture in a greased bowl, oil the top of the dough, and cover the bowl with a wet cloth. Let the dough rise until double (approximately 1½ hours).

Punch down, pinch off pieces of dough the size of a golf ball, flatten. Put a ¼-inch cube of butter on each roll and fold dough over to make a semicircle. Place rolls on flat pan and let rise until double (approximately 1 hour). Bake at 400°F. for 12 minutes or until lightly browned.

2 dozen rolls

Crackers and Cookies

Amaranth Corn Pones

A complete protein snack incorporating traditional Indian foods.

- 1¼ cups cornmeal, finely ground, white or yellow
- ¼ cup amaranth flour
- 2 tablespoons peanut flour
- 2 tablespoons peanut oil
- ¾ cup boiling water
- Parmesan cheese (optional)*

Preheat oven to 325°F. Combining dry ingredients, stir in oil gradually. Add boiling water slowly, mixing with a spoon and finally kneading dough. Keep adding water until dough holds together and is soft.

With your fingers, flatten rounded teaspoon of batter on an oiled cookie sheet. If a cracker is desired, flatten it as thin as possible. If a thicker, larger corn pone is desired, use more dough and don't press it quite as thin. Bake in 325°F. oven for 40 minutes, or until pones are slightly golden around the edges, removing from pan while still warm. Cool on a rack and store in an airtight container.

2 dozen pones

---

*Parmesan cheese can be sprinkled on top of crackers before baking them, if desired.

Amaranth Fritos

½ cup amaranth flour
⅓ cup cornmeal
⅛ teaspoon kelp
½ teaspoon chili powder (more if you like)
⅓ cup boiling water
oil for frying (peanut oil is recommended)
additional kelp for sprinkling

Mix dry ingredients. Pour boiling water over and mix well. Gather into a ball, and when cool, roll out between sheets of plastic wrap to less than ⅛ inch thick. Cut with a sharp knife into triangles or squares. Heat oil for deep frying, at least 2 inches deep. Fry a few at a time. They should rise to the top immediately and puff up into little pillows without burning if your oil is the correct temperature. Keeping a spatula on them for a few seconds as they go in will help them puff. Remove quickly when barely golden. Drain well on a rack and sprinkle with kelp.

Eat within two days or recrisp a few minutes in a low oven.

2 to 3 dozen fritos

Amaranth Tidbits

⅔ cup sifted whole wheat flour
¾ teaspoon baking powder
2 tablespoons amaranth seeds
1 teaspoon vanilla
2 tablespoons honey
1 tablespoon melted butter
2 tablespoons warm water

Sift flour and baking powder together. Add the amaranth seeds, pounded in a mortar or with a kitchen mallet; then add the vanilla, honey, and melted butter. Add the warm water, a tablespoon at a time, just enough to make the dough the consistency of pie crust dough. Place on a lightly buttered cookie sheet without sides.

Cover with floured waxed paper and roll as thin as possible. Remove the paper. Cut the dough into ½-inch squares with a pastry wheel. Do not separate; the tidbits will come apart easily when baked. Bake in a 375°F. oven for 5 to 10 minutes. When light brown, break one open to see if it is done inside. Let cool before breaking into squares.

about 10 dozen tidbits

Am-Ra-Wa Cookies

½ cup vegetable oil
½ cup honey
2 eggs, slightly beaten
1 teaspoon vanilla
½ cup amaranth flour
¾ cup whole wheat pastry flour
½ cup skim milk powder
½ cup raisins
½ cup walnuts, chopped
½ cup wheat germ

Mix oil and honey until well combined, using a whisk. Add eggs and vanilla, blending well. Combine and sift amaranth flour, pastry flour, and skim milk powder. Add raisins to dry mixture, separating them. Add walnuts and wheat germ. Add flour mixture to wet mixture and stir until combined. Drop by teaspoonfuls onto lightly oiled cookie sheet. Bake in preheated oven at 350°F. for 15 minutes.

3 dozen cookies

Animal Crackers

¼ cup amaranth flour
¼ cup raw wheat germ
1 cup oat flour
½ teaspoon curry powder
½ teaspoon kelp
2 teaspoons brewer's yeast (optional)
¼ cup oil (sesame oil is recommended)
⅓ cup boiling water

Mix all dry ingredients together thoroughly. Put oil in a measuring cup, add the boiling water, mix quickly, and pour over the dry ingredients. Stir until the dough is well mixed. Roll out on a floured cloth or board to less than 1/8 inch thick, placing a piece of plastic wrap between dough and rolling pin to prevent sticking. It is important to roll them very evenly so that they bake evenly. Cut out elephants and tigers with zoo cookie cutters and place on an oiled baking sheet. Bake in a 350°F. oven for 8 to 10 minutes. Cool on rack and store in a cool, dry spot, tightly covered. These are great with yogurt cheese dips.

about 3 dozen crackers

Graham Crackers

- 1/4 cup amaranth flour*
- 1 cup whole wheat pastry flour*
- 1/3 cup date sugar (if not available, use 1/4 cup honey and 2 extra tablespoons whole wheat pastry flour)
- 1/4 teaspoon kelp
- 1/2 teaspoon ground ginger
- 1 teaspoon dry yeast
- 1/4 cup warm water
- 1/3 cup butter

Mix all dry ingredients except yeast. Dissolve yeast in warm water. Cut butter into dry ingredients, then add the yeast and water. Gather into a ball, let stand at least 15 minutes if possible, then roll out to 1/8-inch thickness. Cut into squares and bake about 8 minutes on an ungreased cookie sheet in a 350°F. oven. Cool on rack.

The same recipe without the yeast, and using ice water, will make a tender pie shell.

about 2 dozen crackers

---

*1/2 cup of each flour is equally delicious and more nutritious. The proportion depends on how much amaranth you have at hand.

Desserts and Snacks

Amaranth Indian Pudding

- 3 tablespoons amaranth flour
- 2/3 cup cornmeal
- 3 cups milk
- 2/3 cup dark molasses
- 3 tablespoons honey
- 3/4 teaspoon cinnamon
- 3/4 teaspoon nutmeg
- 1/4 cup butter
- 1 cup milk
- yogurt for serving

Combine flours and 3 cups milk. Heat slowly over low heat, stirring. Simmer until thick, about 10 minutes. Add remaining ingredients except 1 cup of milk. Pour into a well-buttered 2-quart baking dish, then pour the remaining cup of milk over top without mixing it in and bake in a very slow oven (275°F.) for 3 to 4 hours, until firm. Serve with thick yogurt.

8 servings

Hikers' Handy Snack — Amaranth Log

- 2/3 cup toasted amaranth flour
- 2 tablespoons honey
- 4 tablespoons peanut butter
- unsweetened shredded coconut

Toast flour slowly in a heavy skillet, stirring constantly to prevent scorching. Mix all ingredients except coconut very well. Shape into a long log, sprinkle with coconut, and slice as needed, or shape into small balls and roll in coconut. Store in the refrigerator.

1 log

Snack Bar

2 tablespoons amaranth flour
3 tablespoons oat flakes
1 tablespoon wheat germ
2 tablespoons raw honey
3 tablespoons chopped pecans
1 tablespoon unsweetened shredded coconut
extra coconut for rolling

Combine all ingredients, except coconut for rolling, until well mixed. Form into a bar approximately 1½ × 4 inches. Roll in coconut and chill. Slice to serve.

1 bar

Amaranth Sprouts

Here is a perfect snack food, ultra-easy to prepare, tasty and superlatively healthy. No cooking, grinding, toasting, or popping is required. The method is very simple. All you need is a pint or quart canning jar and some nylon netting held on by a rubber band.

Soak 2 to 4 tablespoons of seeds for 12 hours, then rinse them twice daily, tilting the jar to drain while sprouting. The sprouts are ready within 1 to 2 days, depending on the final intended use.

For munching and using raw in salads, you can let them grow for 2 days, till they are about ¼ inch long. But for use in baked goods or candies, these are too soggy and seeds that are just barely sprouted for 24 hours (length of seed) are best.

Try sprouts mixed in green salads to add taste and texture contrast; or blended in Indian-type dishes, for instance, cucumber and yogurt salad or cold soup. The flavor is reminiscent of curry and fenugreek. You can also mix them into nut butters (try cashew), cottage cheese, or mashed tofu. They really perk up a sandwich filling. The inherent "spicy" flavor of amaranth self-seasons these sprouts and anything to which they are added.

APPENDIX B

Popped Amaranth Grain

Popping is a tasty and convenient method of preparing the grain. No further cooking is involved and there is no need for a grain or seed mill to grind it. You will find it as, or more, delicious than popped corn. The popping technique is a bit different than corn, however, but not at all difficult.

Lacking the traditional Mexican *comal,* a clay griddle, you can use an ungreased steel wok over medium-high heat. Cast iron works well, too, but the seeds pop out over the stove more readily than when using the wok. Pop small amounts at a time, working quickly to prevent burning and stirring continually with a natural-bristle pastry brush. Remove seeds as soon as they are popped. One tablespoon of seeds will yield about 3 tablespoons popped amaranth. If you find you are getting a very low percentage of "poppers," try sprinkling the seeds with water before popping. If the pan is too hot, the seeds turn black and stick; if too cool, they will not pop.

There are unlimited uses for popped amaranth. It is too small to eat like popcorn, but try it:

in a bowl as a cold breakfast cereal, with milk and honey
mixed with granola or a hot cereal
sprinkled on salads or cooked vegetables
stirred into nut butters
ground in a mortar
with date sugar and cinnamon
sprinkled on toast
sprinkled on yogurt sundaes, even on ice cream
as a breading

Alegrías

(a Mexican confection)

3 tablespoons unsulfured molasses (or honey)
1 cup popped amaranth

Boil the molasses until it "pulls a thread." You may substitute honey for a milder flavor. Mix the syrup into the popped amaranth and knead very well until it forms a ball. Then press the mass onto a well-oiled plate or jelly roll tin until it is quite compact. Cut into squares with a sharp knife while still warm. When it cools, the candy becomes hard and looks something like popcorn balls and sesame seed candy.

Coarsely ground peanuts can be added to the warm mass. Other variations include small amounts of ground pecans or unsweetened coconut.

about 12 small squares

Amaranth Fruit Cookies

- 2 cups whole wheat flour*
- 2 teaspoons skim milk powder
- 4 teaspoons baking powder
- 1½ teaspoons nutmeg
- ½ cup vegetable oil
- ½ cup honey
- ⅓ cup molasses
- 1 egg, well beaten
- 1 cup popped amaranth
- 1 cup chopped, dried fruit (raisins, apricots, etc.)

Combine flour, milk powder, baking powder, and nutmeg. Set aside. Combine oil, honey, molasses, and egg. Gradually blend in the flour mixture. Add the popped amaranth and dried fruit, mixing thoroughly. Drop by teaspoonfuls onto buttered cookie sheets. Batter will spread. Bake for 8 to 10 minutes at 400°F.

5 to 6 dozen cookies

---

*Add ½ cup more flour for a softer, less crunchy cookie.

Amaranth Loaf

- 1 cup popped amaranth
- 1 cup soft whole wheat bread crumbs
- 1 cup chopped peanuts, unsalted raw or roasted
- $\frac{1}{2}$ teaspoon sage
- $\frac{1}{2}$ teaspoon thyme
- pepper to taste
- 2 eggs
- 1 8-ounce can tomato sauce

Butter a $7\frac{3}{8} \times 3\frac{5}{8}$-inch baking pan. Mix popped amaranth, whole wheat bread crumbs, and nuts together. Mix in sage, thyme, and pepper. Add eggs and enough tomato sauce (approximately $\frac{1}{4}$ cup) or cold water to hold the mixture together. Spread into the loaf pan and bake for 45 minutes at 350°F. Heat remaining tomato sauce on medium heat. Cover loaf with warm tomato sauce and serve.

4 servings

Toasted Zucchini

An interesting texture when popped amaranth is used as a breading.

- 3 medium zucchini
- $\frac{1}{2}$ cup popped amaranth*
- 3 tablespoons vegetable oil
- 1 egg, beaten

Cut zucchini into $\frac{1}{4}$-inch slices. Place popped amaranth in a bowl. Heat oil in a large skillet over medium heat. Dip zucchini slices in egg, then in amaranth. Cook in hot oil until golden. Serve piping hot.

4 servings

---

*You can use $\frac{1}{4}$ cup popped amaranth and $\frac{1}{4}$ cup wheat germ instead.

Powdered Amaranth Seasoning

On the way to popping amaranth, you can stop off and prepare a tasty all-purpose table seasoning. Gently toast your seeds on a dry pan until fragrant, but not popping. Quickly remove and cool. Grind these in your seed grinder or crush in a mortar. Pour into a shaker jar and place on the table alongside the traditional seasonings. Sprinkle liberally on any food which needs a little zip. You just might find yourself using less salt. This same seasoning base yields a good soup stock (see recipe for Amaranth Three-Flavor Consomme) and can be brewed as a hot drink.

Cold Amaranth Drinks

The Mexicans grind *Amaranthus hypochondriacus* and mix it with water and syrup to make a pleasant-tasting drink. In India, it is mixed with milk and allowed to ferment. Dr. Campbell of Texas A. & M. University who has visited with Mexican Indians many times, suggested that amaranth flour be mixed with lime as he has seen the natives do to prepare a drink.

Hot Brewed Amaranth

Use about 1 tablespoon of ground toasted amaranth seeds to a cup of boiling water, steep 10 minutes, and pour off the clear liquid. Sweeten with honey. If desired, mix other favorite spices, such as cloves, ginger, cinnamon, or licorice with the powder before steeping.

1 serving

APPENDIX B

Greens Recipes

Soups

Amaranth and Onion Soup

1 pound fresh amaranth or New Zealand spinach, coarsely chopped
1 cup chicken stock
4 tablespoons butter
½ cup spring onions (scallions), thinly sliced, including 2 or 3 inches of the green tops
½ teaspoon finely chopped garlic
5 cups chicken stock
freshly ground black pepper
dash Parmesan cheese

Cook the amaranth or spinach in the cup of chicken stock until tender. Put this mixture in a blender and whirl a few minutes.

Slowly melt the butter in a large, heavy frying pan. Add the scallions and garlic and, over low heat, cook until golden. Combine the scallions, garlic, and blended spinach in a deep saucepan. Add 5 cups of strong chicken stock and a little black pepper.

Without covering the pan, bring the soup to a boil, then reduce the heat and simmer, still uncovered, for 15 or 20 minutes. Shortly before it is done, sprinkle in some Parmesan cheese.

8 cups

Amaranth Three-Flavor Consomme

4 cups boiling water
3 tablespoons toasted and ground amaranth grain
shredded young amaranth greens, a large handful
1 tablespoon tamari soy sauce
2 tablespoons or more popped amaranth

Pour boiling water over ground amaranth. Cover and remove from heat. After 10 minutes, strain and bring to a boil again. Add the greens, remove from heat, stir in tamari soy sauce, and sprinkle with popped amaranth. This is an invigorating, delicate first course.

3 to 4 servings

Curried Amaranth Soup

- 2 potatoes, cubed
- 2 to 3 cups chicken stock
- 1 cup cooked amaranth (2½ cups raw)
- 1 small onion
- 1½ cups milk
- 1 teaspoon curry powder

Cook potatoes in chicken stock until tender. Place potatoes, stock, amaranth, and onion in blender and blend. Add milk and curry powder to taste, and reheat gently, correcting the seasoning.

6 to 8 servings

Mushroom Soup Oriental

- 1 quart chicken or beef stock
- 3 tablespoons tamari soy sauce
- 1 cup diagonally sliced celery
- 1 cup thinly sliced carrots
- ½ cup sliced onions
- 2 cups cooked, chopped, and drained amaranth leaves
- 1 cup sliced mushrooms

In a large saucepan bring stock and tamari soy sauce to a boil. Add celery, carrots, and onions. Reduce heat; cover and simmer for 5 minutes. Add amaranth and mushrooms. Return to boiling point. Reduce heat and simmer for 5 minutes or until vegetables are crisp-tender.

8 to 10 servings

Potato-Amaranth Soup

- 1/4 cup butter
- 1 large onion, chopped fine
- 2 stalks celery, cut fine
- 1/4 cup all purpose flour
- 3 cups water
- 2 cups cubed potatoes
- 1 cup fresh amaranth, chopped fine
- 1/2 cup light cream
- shredded Swiss, Gruyere, or grated Parmesan cheese

Melt butter in a large saucepan. Add onion and celery. Cover. Cook over low heat until tender, about 15 minutes, stirring occasionally. Blend in flour. Add water and potatoes, stirring constantly. Bring to a boil, cover, and simmer about 30 minutes, until potatoes are tender. Stir occasionally. Add amaranth. Simmer, uncovered, 2 minutes. Add cream. Heat thoroughly. Serve hot, sprinkled with cheese.

6 servings

Side Dishes

Amaranth and Onion Sauce

- 1 tablespoon butter
- 2 tablespoons onion, chopped fine
- 1/2 clove garlic, crushed
- 2 cups chopped, fresh amaranth
- 3/4 cup water
- 1 tablespoon cornstarch
- 1/8 teaspoon ground ginger
- 1 tablespoon tamari soy sauce

In skillet, melt butter over medium heat. Add onion and garlic; sauté about 3 minutes, stirring frequently, until lightly browned. Add amaranth and cook, stirring constantly, 2 minutes or until limp. Mix water, cornstarch, ginger, and soy sauce until smooth; stir into amaranth mixture. Bring to boil over medium heat, stirring constantly, and boil 1 minute. Good on meat loaf.

about 1 cup

Amaranth Hush Puppies

- 4 cups fresh amaranth, washed
- 1 cup finely ground cornmeal
- 1 teaspoon baking powder
- 1 egg, slightly beaten
- ½ cup milk
- 2 teaspoons melted bacon or ham fat
- 1 small onion, grated
- corn oil for deep frying

In a large saucepan, cook amaranth until tender. Drain and set aside to cool. Combine cornmeal and baking powder in a large mixing bowl. Add egg, milk, fat, and grated onion and stir to form smooth batter. Chop cooked amaranth fine and add to batter. Drop by teaspoon into hot (365°F.) corn oil. Turn when they rise to the top of the oil and continue cooking until they are golden brown — approximately 5 minutes.

These hush puppies are excellent with fish or barbeque and can be served as snacks for a party.

24 hush puppies

Amaranth, Roman Style

- 2 cups water
- 2 pounds fresh amaranth, washed and drained
- ¼ cup olive oil
- 2 cloves garlic, minced
- 3 tablespoons pine nuts, sliced almonds, pumpkin seeds, or sunflower seeds
- ¼ cup sliced green olives
- ¼ cup sliced black olives
- 1 tablespoon capers
- 2 tomatoes, quartered
- 1 8-ounce can tomato sauce
- 1 teaspoon basil
- ½ teaspoon oregano
- honey to taste

Boil water in large pot. Add amaranth, cover and cook for 5 to 10 minutes until tender, then drain and chop.

Heat oil in a skillet; stir in the garlic and nuts and sauté until golden. Add the olives and capers, mixing until coated. Mix in the amaranth, tomatoes, tomato sauce, basil, oregano, and honey to taste. Heat and serve.

6 servings

Amaranth with Yogurt

2 tablespoons butter
1 medium onion, chopped
1 medium pepper, chopped
1 quart young amaranth greens, washed and chopped
1 cup yogurt
pepper to taste

Melt butter and sauté onion and pepper gently until soft (approximately 10 minutes).

Add amaranth to onion and pepper and cook slowly, covered, for 5 minutes. Do not overcook. Stir in yogurt and add pepper to taste.

4 servings

Chinese-Style Amaranth

1 tablespoon vegetable oil
1 to 2 cloves garlic
1 onion, sliced
1 green pepper, chopped
4 cups amaranth, chopped
1 cup mung bean sprouts
1 tablespoon tamari soy sauce
1 tablespoon water

Heat oil in heavy skillet. Add garlic, onion, and green pepper and stir-fry for a few minutes. Stir in amaranth and sprouts. Mix soy sauce and water and add to vegetables. Cover and cook over medium heat just until amaranth is tender, for 3 to 5 minutes.

4 servings

Corn and Amaranth Skillet

- 4 scallions, chopped
- 2 tablespoons sunflower seed oil
- 1 cup corn kernels cut from the cob
- ½ pound fresh, young amaranth leaves
- kelp or powdered amaranth seasoning, to taste

Sauté scallions in oil. Add corn and amaranth and stir-fry until well coated with oil. Lower heat to minimum, cover closely, and steam 5 minutes. Season with kelp or Powdered Amaranth Seasoning (see recipe in this appendix).

2 servings

Creamy Amaranth Dressing

- 1½ cups steamed amaranth greens
- 1 cup mayonnaise
- 1 small onion
- ½ cup fresh parsley
- 1 teaspoon dried dill seed

Blend all ingredients in a food processor or blender until smooth. Can be served as a dip or a dressing on baked fish.

2½ cups

Fried Amaranth Stems

- amaranth stems, used when they're about 8 to 12 inches tall
- egg, beaten
- flour
- pepper to taste
- vegetable oil for frying

Cut stems into 3-inch lengths. Parboil, then roll in beaten egg and then flour. Add a bit of pepper and fry until browned and tender.

Marinated Amaranth Salad

- ½ cup vegetable oil
- 2 cloves garlic, minced or pressed
- 3 tablespoons white wine vinegar
- 3 tablespoons Dijon mustard
- ¼ teaspoon each pepper and ground cumin
- ½ teaspoon honey
- 2 pounds amaranth greens
- ¼ cup chopped green onion
- 1 or 2 hard-cooked eggs, chopped

Pour oil into a jar or container; add garlic, vinegar, mustard, pepper, cumin, and honey. Cover and shake vigorously to blend. Set aside.

Wash amaranth under running water and drain. Cut into ½-inch strips and steam until tender, 5 to 7 minutes. Cool cooked amaranth to room temperature.

Combine amaranth with green onion. Shake dressing, if necessary, and add to amaranth mixture; stir together. Cover and chill at least 2 hours. Garnish with chopped egg and serve chilled or at room temperature.

6 servings

Salad Dressing

- ½ cup vegetable oil
- ¼ cup vinegar
- 1 tablespoon honey
- 1 tablespoon chives, chopped
- 1 tablespoon amaranth leaves, chopped
- 1 tablespoon amaranth seed, popped

Blend all ingredients in blender, or put in bottle and shake well.

about 1 cup

Scalloped Potatoes

5 potatoes, sliced 1/4 inch thick (leave skins on)
pepper
2 cups cooked amaranth
Mushroom Sauce (see recipe below)
1 cup grated Cheddar cheese
paprika

Cover potatoes with water and bring to a boil. Discard water and cool. Butter a 1 1/2-quart casserole dish and put 1 layer of potato slices in it. Alternate layers of potatoes with layers of amaranth; season with a dash of pepper.

Pour mushroom sauce over the top layer, spreading it evenly. Cover with grated cheese and sprinkle with paprika. Bake in 350°F. oven for 1 hour.

6 to 8 servings

Mushroom Sauce

12 mushrooms, sliced
2 tablespoons butter
1 1/2 tablespoons cornstarch
1 cup milk
dash of pepper

Sauté mushrooms in 1 tablespoon of the butter. Melt the second tablespoon of butter in a heavy saucepan over low heat. Add cornstarch and mix well. Slowly add milk, stirring constantly. Bring mixture to boil and cook for 2 minutes, stirring constantly. Season to taste with pepper and add mushrooms.

1 cup

Steamed Amaranth

1 pound amaranth
2 tablespoons vegetable oil
1 clove garlic, minced
1 onion, chopped
3 hard-cooked eggs, diced (optional)

Thoroughly wash and coarsely chop amaranth. Heat oil, then brown garlic and onion. Add amaranth and stir thoroughly. Cover and allow to steam in its own water — approximately 5 to 8 minutes.

For variation you can add 3 diced, hard-cooked eggs before serving.

4 servings

Stir-Fried Amaranth with Peanuts

- 2 pounds fresh amaranth
- 1/4 cup water
- 6 tablespoons vegetable oil
- pinch of turmeric
- 3/4 teaspoon ground cumin
- 1/2 teaspoon ground coriander
- 1/4 cup chopped peanuts

Wash and chop amaranth fine. Boil 1/4 cup water. Add chopped amaranth, cover, and cook about 6 minutes or until tender. Drain well. Heat vegetable oil in a skillet and add turmeric, ground cumin, and coriander.

Add amaranth and continue cooking and stirring, over medium-low heat, about 5 more minutes. Stir in peanuts, and cook slowly for about 2 minutes. Remove from heat and serve.

4 to 6 servings

Summer Squash with Amaranth

- 1 onion, chopped
- 3 tablespoons soy oil
- 2 small yellow squash or zucchini, cubed
- 1/2 cup cornmeal
- 1 1/2 cups water
- 2 cups amaranth greens, chopped

In large skillet, sauté onion in oil until golden. Stir in squash, then cornmeal; add water, and stir again. Cover and cook on low heat for 20 minutes, stirring several times. Add amaranth and cook 5 more minutes.

6 servings

Main Dishes

Amaranth and Cream Cheese Casserole

- 4½ quarts fresh amaranth greens
- 2 cups water
- 4 3-ounce packages cream cheese
- 3 tablespoons butter
- ½ cup yogurt
- pepper and nutmeg to taste
- ⅔ cup fresh, grated Parmesan cheese

Wash and steam amaranth in 2 cups water for 5 minutes. Squeeze the amaranth dry as possible and spread evenly over bottom of buttered, 8 × 10-inch pan. Beat cream cheese and butter together until it is smooth and gradually add the yogurt. Spread over the amaranth mixture, dust with pepper and nutmeg. Cover with Parmesan cheese. Cover with foil and bake at 375°F. for 40 minutes.

8 to 10 servings

Amaranth and Rice

- 2 pounds amaranth
- ¼ cup olive oil
- 1 small onion, chopped
- 4 tomatoes, chopped
- ½ teaspoon pepper
- 1 tablespoon dill weed
- 2 cups water
- 1 cup raw brown rice

Wash amaranth, slice into small pieces, and sauté in oil along with onion. When amaranth is limp add tomatoes, pepper, dill, and water. Bring to a boil and stir in rice; reduce heat, cover, and simmer until rice is tender, about 30 minutes.

8 servings

Amaranth Au Gratin

- 1 pound fresh amaranth
- 1 pound ricotta
- 1 egg, beaten
- ½ cup freshly grated Parmesan cheese
- pepper

Cook amaranth for 7 minutes in boiling water in a covered pot until tender. Drain amaranth. Combine with ricotta, egg, and ¼ cup of the grated Parmesan cheese. Pour into a 2-quart casserole dish. Sprinkle remaining cheese on top. Bake at 350°F. oven for 30 minutes or until cheese is golden brown.

6 servings

Amaranth Foo Yung

- 2 tablespoons safflower oil
- 2 onions, chopped fine
- 2 cups fresh amaranth, chopped
- 6 eggs, slightly beaten

Pour oil into skillet and heat at medium temperature. Mix remaining ingredients. Fry about 2 tablespoons of amaranth-egg mixture at a time; do not stir, but cook until the pancake is lightly browned on both sides. Continue frying pancakes, 1 at a time. The pancakes can be kept warm in the oven in a flat pan at low temperature.

4 servings

Amaranth Frittata

- 4 green onions, sliced
- 1 tablespoon butter
- 8 eggs
- 1 cup half-and-half
- 2 cups shredded amaranth
- 2½ cups (10 ounces) shredded Cheddar cheese
- snipped parsley

Cook and stir onions in butter until tender. Beat eggs with half-and-half. Stir in onions, amaranth, and 2 cups of the cheese. Pour into buttered 11¾ × 7½-inch baking dish. Bake at 350°F. until mixture is set and top is light brown, 25 to 30 minutes. Sprinkle frittata with remaining ½ cup cheese. Bake until cheese melts. Cut frittata into rectangles and serve.

6 to 8 servings

Amaranth and Lamb

- 2 pounds boneless, lean lamb, cut in 2-inch cubes
- 4 tablespoons butter
- 2 medium onions, sliced very thin
- 2 teaspoons ground coriander
- 2 teaspoons ground ginger
- 1 teaspoon turmeric
- ½ teaspoon chili powder
- 1 pound fresh amaranth, finely shredded
- 2 tablespoons plain yogurt
- 1 tablespoon mustard seed
- dash of crumbled, dried thyme
- additional yogurt, for sauce

In a large skillet with cover, quickly and lightly brown lamb in 2 tablespoons of the butter remove from pan and set aside.

Melt remaining butter in skillet, add onions, and sauté just until limp. Return meat to skillet with next 4 ingredients; sauté for 15 minutes over medium heat, stirring occasionally.

Sprinkle with amaranth, 2 tablespoons yogurt, mustard seed, and thyme; mix well. Cover and cook over low heat 45 minutes, or until lamb is tender, the amaranth like a puree, and the liquid completely cooked down. (During cooking, stir occasionally and add water, a little at a time, as necessary.) Pass additional yogurt as a sauce.

8 servings

Amaranth Lasagna

1 pound fresh amaranth, washed and drained
2 cloves garlic, minced
3 tablespoons dried parsley
1 tablespoon basil
1 teaspoon oregano
½ cup wheat germ
3 8-ounce cans tomato sauce
1 6-ounce can tomato paste
9 whole wheat lasagna noodles
1 teaspoon vegetable oil
1 pound ricotta cheese
dash pepper
½ cup grated Parmesan cheese
½ pound mozzarella cheese, sliced

Boil a small amount of water and add amaranth. Cover and cook until tender. Drain.

Place amaranth in blender and add garlic, 1 tablespoon parsley, basil, and oregano. Blend carefully until mixed but not a liquid. Now add wheat germ, 2 8-ounce cans tomato sauce, and tomato paste. Stir to blend.

Cook lasagna noodles according to package directions. Add oil to cooking water so noodles won't stick together.

Mix ricotta cheese, pepper, and remaining 2 tablespoons of parsley.

Butter the bottom of a 9 × 13-inch baking pan. Put in 3 lasagna noodles next to each other as a first layer. Then put in a layer of the amaranth mixture, a layer of ricotta, shake a little Parmesan over the ricotta, and then some slices of mozzarella. Start again with the noodles and keep doing this, ending with the amaranth mixture. Cover with remaining cheese and a can of tomato sauce.

Bake for 30 minutes at 375°F.

10 to 12 servings

Amaranth Pasta

For those of you who love green spinach noodles, here's a terrific variation.

2 cups whole wheat pastry flour
2 eggs, beaten
1 to 2 tablespoons water
1 tablespoon olive oil
¼ pound fresh amaranth, steamed and chopped fine

Reserve ¼ cup of the flour for dusting the board. Mound the flour on your work surface and make a well in the center. Pour in the beaten eggs, water, oil, and amaranth. Slowly combine dry and liquid ingredients into a smooth ball. Knead briefly on a floured board or cloth, cover, and let rest at least ½ hour. Then roll out the dough, ¼ at a time, as thin as possible. Dry 10 minutes, then cut into thin strips. Use immediately or dry thoroughly for future meals. The amaranth lends a bright green color and creaminess to your pasta.

4 to 6 servings

Amaranth Quiche

Pie Crust:
1 cup plus 2 tablespoons unbleached flour
⅓ cup corn oil
2 tablespoons cold water

Filling:
4 cups amaranth leaves, sliced
¼ cup chopped onion
2 tablespoons butter
1 tablespoon whole wheat flour
3 eggs
1 cup milk
2 cups natural Swiss cheese, shredded

Blend flour and corn oil thoroughly with a fork. Sprinkle all of the water over the mixture and mix well. Press dough firmly into a ball with hands. If dough is too dry, add 1 or 2 tablespoons more oil. Roll out dough and place in 9-inch pie pan. Partially bake pie crust at 450°F. for 5 to 7 minutes. Remove and set aside.

Sauté amaranth and onion in butter for 10 minutes. Add flour. Beat eggs and milk and add half to the amaranth. Put half of the cheese in the pie crust. Top with the amaranth mixture and add the remaining egg-milk mixture. Sprinkle remaining cheese over the top. Bake in a 325°F. oven for 40 to 45 minutes or until knife blade comes out clean. Let the quiche stand for 10 minutes to set before serving.

6 servings

Amaranth Roll-Ups

- 2 tablespoons butter
- 1/3 cup chopped onion
- 3 tablespoons cornstarch
- dash pepper
- dash ground thyme
- 1 1/2 cups milk
- 1/3 cup mayonnaise
- 4 cups chopped amaranth greens
- 2 7-ounce cans tuna, drained
- 6 whole wheat lasagna noodles; cooked, drained, and cut in half
- 2 tablespoons grated Parmesan cheese

Melt butter in saucepan; sauté onion until tender. Add next 5 ingredients. Heat until thickened; stir in amaranth. Mix 1/3 cup sauce with tuna. Spoon 2 tablespoons tuna mixture on each noodle and roll up. Pour half of remaining sauce into baking dish; place roll-ups on top and pour on rest of sauce. Sprinkle with cheese. Bake in 325°F. oven for 20 minutes.

12 rolls

Amaranth Scrambled Eggs

- 2 cups young, tender amaranth leaves
- 3 slices bacon
- 2 eggs
- pepper

Steam amaranth leaves in a small amount of boiling water in a covered pot until tender. Drain and chop. Fry bacon in a skillet until crisp. Cut into small pieces and add drained greens. Break eggs into mixture and scramble until eggs are well done. Season with a dash of pepper.

1 serving

Amaranth Soufflé

- 7 tablespoons butter
- 7 tablespoons flour
- 1¾ cups milk
- white pepper to taste
- ¼ cup onion, chopped
- ¼ cup green pepper, chopped
- 1 pound amaranth, steamed
- ¼ pound mushrooms, thinly sliced
- ½ cup Cheddar cheese, grated
- 4 eggs, separated
- ¼ teaspoon basil
- ¼ teaspoon rosemary, crushed

Lightly butter a 2-quart soufflé dish and fit it with a 6-inch-wide band of waxed paper, doubled and buttered, to form a standing collar extending 2 inches above the rim.

In a large saucepan melt 6 tablespoons butter, stir in flour, and cook over low heat, stirring, for 3 minutes. Remove the pan from the heat and add milk, whisking vigorously until the mixture is thick and smooth. Add white pepper to taste. Heat through.

In a skillet, sauté onion and green pepper in 1 tablespoon butter until the onion is lightly browned, and add the vegetables to the sauce. Stir in amaranth; mushrooms; Cheddar cheese; 4 egg yolks, lightly beaten; basil; rosemary; and white pepper. In a bowl, beat 4 egg whites until they hold stiff peaks, and fold them into the vegetable mixture. Spoon the mixture into the soufflé dish and bake it in the center of a preheated moderate oven (350°F.) for 55 to 60 minutes, or until it is puffed and brown. Remove the collar and serve the soufflé immediately.

4 to 6 servings

Amaranth Spaghetti Ring Florentine

8 ounces whole wheat spaghetti
2 quarts chopped amaranth
½ cup chopped onion
½ cup grated Romano cheese
1 2-ounce jar pimento, chopped and drained
6 tablespoons butter
2 eggs, slightly beaten
3 cups sliced, fresh mushrooms
4 cups marinara or spaghetti sauce

Cook spaghetti and drain. Steam amaranth with chopped onion and drain. Combine well-drained spaghetti and amaranth, cheese, pimento, 4 tablespoons of the butter, and eggs. Mix well. Turn into a greased 6½-cup ring mold, cover and bake at 375°F. for 25 minutes.

Cook mushrooms in remaining butter and add sauce. Heat through.

Cool ring upside down for 10 minutes before serving. Remove mold and serve with sauce on the side.

6 to 8 servings

Amaranth-Tofu Stir-Fry

1 tablespoon cornstarch
½ cup cold water
2 tablespoons tamari soy sauce
1 clove garlic, minced
2 tablespoons vegetable oil
1½ cups chopped onion
2 cups chopped amaranth, steamed
1 6-ounce can water chestnuts, sliced (1 cup)
1½ cups mung bean sprouts
1 cup cubed tofu
½ cup toasted, slivered, blanched almonds

Combine cornstarch with water and soy sauce.

Sauté garlic in oil. Add onion and stir-fry until tender. Add amaranth and water chestnuts; stir-fry a few minutes.

Add cornstarch mixture. When sauce has thickened add sprouts and tofu, and toss lightly to cover with sauce.

Arrange on platter and sprinkle almonds on top.

4 servings

Amaranth with Noodles

- 1½ cups chopped, cooked amaranth
- 1½ cups cooked whole wheat noodles
- ¼ teaspoon pepper
- 1 cup grated cheese
- 1 cup milk

Put alternate layers of amaranth and noodles in a greased baking dish, sprinkle pepper and cheese over each layer, and top with a layer of noodles. Pour milk over all and bake in a moderate oven (350°F.) for 45 minutes.

2 to 3 servings

Baked Amaranth with Cheese

- 2 pounds amaranth
- 4 tablespoons butter
- 3 tablespoons all-purpose flour
- 1¼ cups milk or half-and-half
- 2 cups shredded Swiss cheese
- pepper
- about ¼ teaspoon ground nutmeg

Wash amaranth under running water and drain. Stack the leaves and cut crosswise into ½-inch strips; steam until tender, 5 to 7 minutes.

To make sauce, melt the butter in a small saucepan over medium heat and stir in flour; cook, stirring, until bubbly. Gradually stir in milk, and cook, stirring until mixture thickens, about 10 minutes. Remove from heat and stir in ¾ cup of the cheese; add pepper to taste.

Line the bottom of an ungreased 2-quart baking dish with

a few of the cooked leaves and spoon about $\frac{1}{3}$ of the sauce on top. Repeat with remaining leaves and sauce to form 3 layers. Top generously with remaining cheese and sprinkle with nutmeg. If made ahead, cover and chill.

Bake, uncovered, in a 350°F. oven for 20 minutes (25 minutes if chilled), until cheese melts and amaranth is heated through. Then place dish under a broiler 2 to 3 minutes to lightly brown cheese.

4 servings

Green Rice Casserole

- 1 cup shredded Cheddar cheese
- 1 tablespoon butter
- 2 cups cooked brown rice
- 2 tablespoons chopped onion
- 2 tablespoons minced parsley
- 1 cup steamed amaranth leaves ($2\frac{1}{2}$ cups fresh)
- $1\frac{1}{2}$ cups milk
- 2 eggs, slightly beaten

Stir cheese and butter into hot rice. Add onion, parsley, and amaranth. Stir milk into eggs. Combine with rice mixture. Pour into oiled casserole. Bake at 325°F. till set, about 35 minutes.

6 servings

Green Vegetable Casserole

4 cups water
1 pound amaranth, washed, drained, and chopped
1 pound green cabbage (coarse outer leaves removed), washed, quartered, and sliced
2 leeks, trimmed, washed, and sliced
¼ cup oil, soy or safflower
3 small zucchini, trimmed, washed, and cut into ¼-inch slices
2 garlic cloves, crushed
1 cup grated Cheddar cheese
3 eggs, lightly beaten
¼ cup light cream
½ teaspoon white pepper
¼ teaspoon cayenne pepper
½ cup raw wheat germ

Bring 2 cups of the water to a boil. Reduce the heat to moderate, add amaranth, cover, and cook the amaranth until it is just tender.

Drain the amaranth in a colander and set the amaranth aside in a large mixing bowl.

Meanwhile, put the cabbage and leeks into a large saucepan and add 2 cups water. Place the pan over moderately high heat and bring the water to a boil. Reduce the heat to moderate and cook the vegetables for 10 minutes. Remove the pan from the heat and drain the vegetables in a colander. Add the cabbage and leeks to the amaranth in the mixing bowl.

In large frying pan, heat the oil over moderate heat. When the oil is hot, add the zucchini and the garlic. Fry them, turning occasionally, for 8 to 10 minutes, or until the zucchini is lightly browned on both sides. With a slotted spoon, transfer the zucchini to the amaranth mixture. Place the amaranth mixture in a greased 2-quart casserole. Mix ½ cup of the grated cheese into the vegetables.

In small mixing bowl, beat the eggs and cream together with a fork. Beat in the pepper and cayenne. Pour the egg mixture over the vegetables in the baking dish.

Sprinkle the remaining ½ cup cheese and the wheat germ over the vegetables. Bake for 30 to 35 minutes at 350°F. or until the topping is melted and brown.

6 servings

Rolled Amaranth Soufflé Stuffed with Mushrooms

- bread crumbs
- 2½ pounds fresh amaranth
- pepper to taste
- ¼ teaspoon nutmeg
- 6 tablespoons butter
- 4 eggs, separated
- 4 tablespoons Parmesan cheese
- ½ pound mushrooms, sliced and sautéed
- ½ cup grated Swiss cheese
- paprika

Butter a 10½ × 15½ × 1-inch baking pan and line with buttered foil or waxed paper. Sprinkle with bread crumbs.

Wash and drain amaranth. Cook amaranth in a little water about 5 minutes, until only slightly done. Drain and cool. Chop fine and drain again, pressing out the water through a colander. Put in a bowl and season with pepper and nutmeg. Stir in 6 tablespoons butter. Beat in egg yolks, one at a time, blending after each one.

Beat egg whites until they form soft peaks; don't overbeat. Fold whites into amaranth mixture. Spoon mixture carefully into prepared pan and sprinkle with Parmesan cheese. Mixture should be ½ to ¾ inch high. Bake at 350°F. for 12 to 15 minutes. Soufflé is done when the sides pull away from the pan.

Remove from oven, and let stand a few minutes. Then cover with buttered foil, and invert onto foil. Remove pan and peel off foil gently, using a long spatula if it sticks a little. Fill soufflé with mushrooms and roll, starting at narrow end. Sprinkle top of roll with grated Swiss cheese. Return to oven and allow cheese to melt a little, then sprinkle with paprika. Serve hot.

8 to 10 servings

Spanakopita
(Amaranth Pie)

Pie Crust:
- ½ cup butter
- 2 cups whole wheat flour
- 4 to 6 tablespoons ice water

Filling:
- 2 pounds amaranth, chopped
- 1 onion, chopped
- vegetable oil
- 3 eggs
- 1 cup milk
- 1 cup cottage cheese
- ½ teaspoon oregano
- pepper
- 3 tablespoons butter, melted
- grated Parmesan cheese, or feta cheese (optional)

For pie crust, cut butter into flour. Toss with the water until it forms a ball. Divide in half. Roll half out and place in a 9-inch pie pan. Reserve the other half.

For filling, sauté the amaranth and the onion in a bit of oil. Spread on the bottom of the pie crust. Mix the eggs, milk, cottage cheese, oregano, and pepper to taste. Pour over the amaranth.

Bring the edges of the crust over this and drizzle butter on the edges and filling. Roll out the top crust and place on the pie, drizzle more butter on the edges and top and fold over any dough hanging over the pie tin. Slash the top in several places and bake 30 to 45 minutes at 375°F.

Parmesan cheese is good sprinkled on the filling before the top crust, and feta cheese is marvelous crumbled in with the cottage cheese.

6 servings

*Leaf Protein Concentrate**

Though green leaves are the earth's most plentiful source of protein, they also contain lots of water and fiber, and have a

---

*Excerpted from *Unusual Vegetables,* ed. Anne Halpin (Emmaus, Pa.: Rodale Press, 1978).

protein-binding structure. All of this means that it's impossible for humans to get a substantial amount of protein from leaves eaten straight off the plant. Researchers have found, though, that when leaves have at least 76 percent water and at least 18 percent protein on a dry weight basis, as the amaranths do, it is possible and worthwhile to make them into a concentrated food tremendously rich in protein, vitamin A, calcium, and iron. This is done by breaking down the structure and fiber of the leaves and then pressing out the water. The pleasantly aromatic, mild-tasting leaf protein concentrate (LPC) gotten this way has a crumbly texture much like dry cottage cheese. Although it takes rather a lot of effort to produce LPC in sizable amounts, the product is very versatile, and may prove especially useful to vegetarians who eat no animal products whatever. Containing up to 65 percent protein, LPC may be blended into stews, sauces, mashed potatoes, and many other dishes.

Amaranth yields a more palatable leaf concentrate than alfalfa or soybeans, so if you are a culinary adventurer, you might want to try converting your mature amaranth leaves into LPC. For best results, harvest the leaves before the plants flower, for those gathered later when you harvest the grain are likely to be bitter.

The yield of this recipe is low, and for regular use you will need to multiply it. But for the purpose of experiment, you'll want to start small. To makc 3 grams (1/10 ouncc) of this ncw food, gradually add ¼ pound of washed amaranth leaves to 1 pint of water in a blender, liquefying as you go. (If your water tends to be alkaline, add a little vinegar.) When all the leaves have been added, continue to blend at high speed for another 2 minutes. Then pour the mixture through a fine cloth spread over a colander or sieve, catching the liquid and squeezing the cloth to get as much of it as possible. Next, heat this liquid quickly to 176°F., stirring all the while. When the mixture reaches the right temperature, allow it to stand, off the heat, until the solid protein starts to separate from the liquid. Then pour the mixture through a fine cloth, allowing the liquid to drain off, and squeeze the solid portion lightly.

While the curd is still in the cloth (in the colander), rinse it with water, stirring lightly. Allow it to drain again, then squeeze it with your hands or place the bag in a potato ricer. When you remove the pressed curd from the cloth, you can use it right away, freeze it, or store it for up to 2 weeks in the refrigerator. You can sneak this green protein into Greek spinach pie or a vegetable casserole. Or add it to peanut or sesame butter and roll into small balls to eat as a snack.

LPC is very valuable in pureed baby foods. Add a tiny bit at first, then increase the amount as your baby becomes accustomed to the taste. Here are two ideas from foreign cuisines.

Guacamole

- 2 medium-size avocados
- ½ medium-size onion, chopped fine
- 4 fresh coriander sprigs, leaves only, chopped fine (or substitute parsley if coriander is not available)
- 1 to 2 fresh green chilies, chopped fine (or substitute a few drops of Tabasco sauce)
- ½ to 1 teaspoon LPC (see above)
- 1 large fresh tomato, peeled and chopped

To be authentically Mexican, the avocados must be mashed with a mortar and pestle, but a wooden spoon or fork will also do. Do not put through an electric blender. Mash the avocados, then grind half the onion, half the herbs, and the chilies to a paste, and mix into the avocados along with the LPC. Add the tomato and remaining onion and herbs. Use immediately or cover tightly with plastic wrap to prevent discoloration. Serve as a dip with Amaranth Fritos (in the grain recipes above).

6 servings

Irish Colcannon

Rich in complete plant protein, calcium, iron, and potassium.

- 4 medium-size potatoes, scrubbed but not peeled
- 8 scallions, sliced
- 1 tablespoon butter
- 3 cups chopped fresh amaranth
- about ¼ cup water
- freshly ground pepper
- 2 tablespoons heavy cream or yogurt
- 1 teaspoon LPC (see above)
- 2 tablespoons chopped parsley

Dice potatoes and cook in a tiny amount of water until just tender. Sauté the scallions in butter, add the amaranth, cover and steam with about ¼ cup water until barely tender. Puree the potatoes, season with pepper and cream or yogurt. Stir in the LPC very thoroughly, then mix in the amaranth and parsley. Spread the mixture in a buttered baking dish and heat through in a medium oven (350°F.) until the top just begins to color.

4 to 6 servings

Baby Foods

Amaranth in any form is a valuable food for a baby when he or she begins to eat solids. Although it is new in this country and feeding tests have not yet been performed, amaranth grain could replace milk in the diet of a weaned baby or young child who is allergic to milk and/or soybeans. It is certainly a practical baby food, being nonperishable. Once harvested, it may be stored until the following harvest and used as needed. Amaranth baby foods are also extremely valuable for the elderly or convalescent when dentures are a problem and few foods are well tolerated. Three tablespoons of amaranth flour contain more protein and assimilable calcium than ½ cup of milk.

Instant Cream of Amaranth Cereal

3 tablespoons finely ground amaranth, toasted very lightly (if desired)*

Stir in just enough boiling water to moisten ground amaranth. Cover and let soften at least 5 minutes. A bit of tahini (sesame butter) may be stirred in for extra smoothness and nutrition. Milk or honey may be added to taste for the elderly.

1 serving

Pureed Greens

4 ounces fresh amaranth greens

Chop greens fine and steam them in a very small amount of water for five minutes. Uncover pan and evaporate any excess moisture. Puree in food mill or food processor. Serve to baby as is, or combined with another vegetable. Butter or oil may be added. Freshly chopped parsley and other green herbs pick up the taste.

1 serving

---

*Toasting improves the flavor, but is somewhat destructive to vitamins. It is certainly not necessary.

APPENDIX B

Recipe Credits

Most of the recipes in this appendix were created by Anita Hirsch, Susan Asanovic, Beth Correll, and Diana Reitnauer in Rodale's Experimental Kitchen. However, a number of *Organic Gardening* Reader-Researchers supplied us with their favorite recipes, and we would like to thank the following for their recipe contributions:

Evelyn Blair
Irene Brackman
Anna Belle Cole
Mr. and Mrs. Milt Cunningham
Dave and Bonnie Fisher
Mrs. Roy R. Fox
Tina Hoffman
John and Bette Hope
Beatrice Trum Hunter
Robert B. Johnstone
Mrs. Ruth Kidwell
Julie Lacitignola
Monica Mashensic
Sara Miller
Edna Newcomb
Louise Riotte
Diane Sciscente
Mrs. Larry Vail
Jane Young

Appendix C
Nutritional Charts

The figures represented here are not fixed values. Depending on growing conditions and the particular strain being analyzed, the values can vary from as much as 25 to 50 percent. However, the figures in these charts can be used effectively to compare the relative food values of the grains and vegetables listed.

FOOD COMPOSITION*
Vitamins in 100 Grams

	Vitamin A (IU)	**Thiamine (mg.)**	**Riboflavin (mg.)**	**Niacin (mg.)**	**Ascorbic Acid—Vitamin C (mg.)**
Amaranth grain	0	0.14	0.32	1.0	3
Corn	70	0.43	0.10	1.9	Trace
Rice	0	0.34	0.05	4.7	0
Soybeans	80	1.1	0.31	2.2	.....
Wheat (hard red spring)	0	0.57	0.12	4.3	0

*Compiled by Joel Elias

COMPOSITION OF RAW GREENS

Selected Nutrients in 100 Grams

	Moisture (%)	Protein (gr.)	Calcium (mg.)	Phosphorus (mg.)	Iron (mg.)	Potassium (mg.)	Vit. A (IU)	Thiamine (mg.)	Riboflavin (mg.)	Niacin (mg.)	Ascorbic Acid—Vitamin C (mg.)
Amaranth	86.9	3.5	267	67	3.9	411	6,100	0.08	0.16	1.4	80
Beet greens	90.9	2.2	119	40	3.3	570	6,100	0.10	0.22	0.4	30
Chard	91.1	2.4	88	39	3.2	550	6,500	0.06	0.17	0.5	32
Collards	85.3	4.8	250	82	1.5	450	9,300	0.16	0.31	1.7	152
Kale	87.5	4.2	179	73	2.2	378	8,900	.....	.....	.....	125
Spinach	90.7	3.2	93	51	3.1	470	8,100	0.10	0.20	0.6	51

Source: USDA. 1963. *Composition of Foods.* Agriculture Handbook No. 8. Washington, D.C.: U.S. Government Printing Office.

COMPOSITION OF GRAINS
Selected Nutrients in 100 Grams

	Food Energy (cal.)	Moisture (%)	Protein (gr.)	Fat (gr.)	Total Carbohydrate inc. Fiber (gr.)	Fiber (gr.)	Ash (mg.)	Calcium (mg.)	Phosphorus (mg.)
Amaranth, whole grain (*A. hypochondriacus*)	391	9.35	15.31	7.12	63.1	2.89	2.61	490	455
Pigweed, whole grain (*A. retroflexus*)	426	10.93	15.84	9.46	63.1	10.75	2.38	490	455
Buckwheat, whole grain	335	11	11.7	2.4	72.9	9.9	2	114	282
Cornmeal, white or yellow, whole ground, unbolted	355	12	9.2	3.9	73.7	1.6	1.2	20	256
Rye grain	334	11	12.1	1.7	73.4	2	1.8	(38)	376
Soy flour, low fat	356	8	43.4	6.7	36.6	2.5	5.3	263	634
Whole wheat flour from hard wheats	333	......	13.3	2	71	2.3	1.7	41	372

Sources: USDA. *Composition of Foods.* Agriculture Handbook No. 8. Food Composition Table for Use in Africa.

APPENDIX C

PROTEIN CONTENT*
(range)

Grain Variety	% Protein
Amaranth	13.6–18
Barley	9.5–17
Corn	9.4–14.2
Rice	7.5
Rye	9.4–14
Wheat	14.0–17

*Compiled by Joseph Senft

Protein Quality

The quality of a protein depends upon its content of essential amino acids which are the first 8 items across the top of the table. Whole egg protein has been determined to be the best protein on the basis of its utilization by animals, so the quality of other proteins can then be determined by comparing its essential amino acid content to that of whole egg. For example, *Amaranthus edulis* has a better lysine content than does *A. hypochondriacus* because the amount of lysine in whole egg is closer to the amount in *A. edulis* than to the amount in *A. hypochondriacus.*

In the same respect, the ratio of 1 amino acid to another is best in whole egg, and one can evaluate the ratios of amino acids in any other food by comparing them to the same ratios in whole egg. For example, the ratio of methionine to threonine in *A. edulis* is better than the ratio of methionine to threonine in *A. hypochondriacus* because it is closer to the ratio of methionine to threonine in whole egg.

You'll note from the table that common grain proteins are relatively deficient in lysine and methionine whereas amaranth has relatively high amounts of these.

AMINO ACID

Grams/G Protein (16 Grams N)

Protein Source	TRY	MET	THR	PHA	LYS	ILE	VAL	LEU	HIS	CYS	TYR	ARG
Whole egg[1]	1.65	3.14	4.98	5.78	6.40	6.60	7.42	8.80	2.41	2.34	4.30	6.56
Amaranth												
A. edulis (New Zealand)[2]	.....	2.3	3.7	4.3	6.2	3.9	4.4	5.9	2.8	1.6	4.1	10.6
A. hypochondriacus[3]	0.18	2	3.2	3.8	5	3.3	3.8	5.2	2.5	2	3.2	8.1
A. cruentus (India)[4]	0.9	2.5	4.3	4.7	8.2	6.9	6	8	2.9	2.9	.....	14.7
A. retroflexus (U.S.)[5]	.....	1.8	3.2	3.5	4.6	3.3	3.7	5.1	2.3	.....	3.3	7.5
200 *angiosperm* seed[5] (average of 200 flowering plants)	.....	1.47	3.27	3.83	4.47	3.56	4.43	6.03	2.27	.....	2.85	8.28
Corn[1]	0.61	1.86	3.98	1.29	2.88	4.62	5.10	12.96	2.06	1.30	6.11	3.52
Wheat[1]	1.15	1.42	2.69	4.61	2.56	4.05	4.32	6.26	1.90	2.05	3.49	4.46
Rice[1]	1.02	1.71	3.73	3.66	3.75	4.46	6.66	8.19	1.6	1.30	4.35	5.40
Soy[4]	1.2	1.1	3.9	4.8	6.6	5.8	5.2	8.6	2.5	1.2	.....	7
Beef[1]	1.17	2.48	4.42	4.11	8.74	5.23	5.55	8.19	3.47	1:26	3.39	6.45

1. M. L. Orr and B. K. Watt, 1957 (1968). *Amino Acid Content of Foods.* USDA. Washington, D.C.: U.S. Government Printing Office.
2. W. J. S. Downton, *Amaranthus edulis. World Crops* 25:20.
3. R. M. Saunders, 1978. Unpublished data.
4. C. E. Smith et al., 1969. Seed Protein Sources. *Economic Botany* 13:132.
5. C. H. Van Etten et al., Amino Acid Composition of Seeds from 200 Angiospermous Plant Species. *Journal of Agricultural and Food Chemistry* 11:399-410.

Appendix D
Catalog of Major Amaranth Species

Amaranths are divided into 4 groups: those identified as grain types, vegetable types, ornamentals, and weeds. There is much variation within these 4 groups. The following catalog briefly identifies and illustrates the primary species that are most commonly available.

Amaranthus hypochondriacus

This most common and heaviest-yielding of all the amaranths originated in Mexico and Central America. And like all amaranths, it is an annual herb. It is characterized by a single, erect, large flower head composed of thick fingerlike projections (spikes). Sometimes these appear as several thick spikes that are over 1 foot long or as groups of shorter spikes forming a dense flower head. The flowers can be red, green, or marbled red and green. The leaves are elliptical. Height varies from 5 to 8 feet.

Use:

The seed is white and it is used as grain — popped, ground, sprouted, or cooked as cereal. The leaves are sometimes eaten when very young. As an ornamental developed in England, *A. hypochondriacus* is known as prince's-feather.

Seed Source:

Seeds of an improved selection of *A. hypochondriacus* are

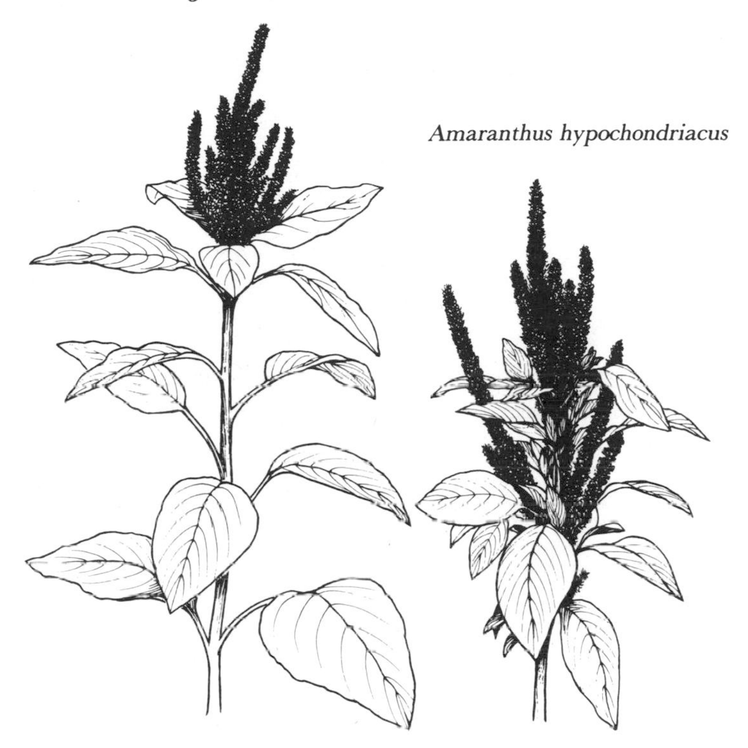
Amaranthus hypochondriacus

available from Gurney Seed and Nursery Co., Yankton, SD 55078; and Johnny's Selected Seeds, Albion, ME 04910.

Amaranthus cruentus

The flower head is composed of loosely arranged spikes on which the flower clusters on the spike are smaller than those of *A. hypochondriacus* and thus give the impression of an open flower head. The flowers of *A. cruentus* are generally yellow green, but occasionally red. Leaves are thinly elliptical and are on a relatively long petiole (the stem of the leaf). The plant height ranges from 4 to 7 feet.

Use:

The black seeds of *A. cruentus* have been used as a grain

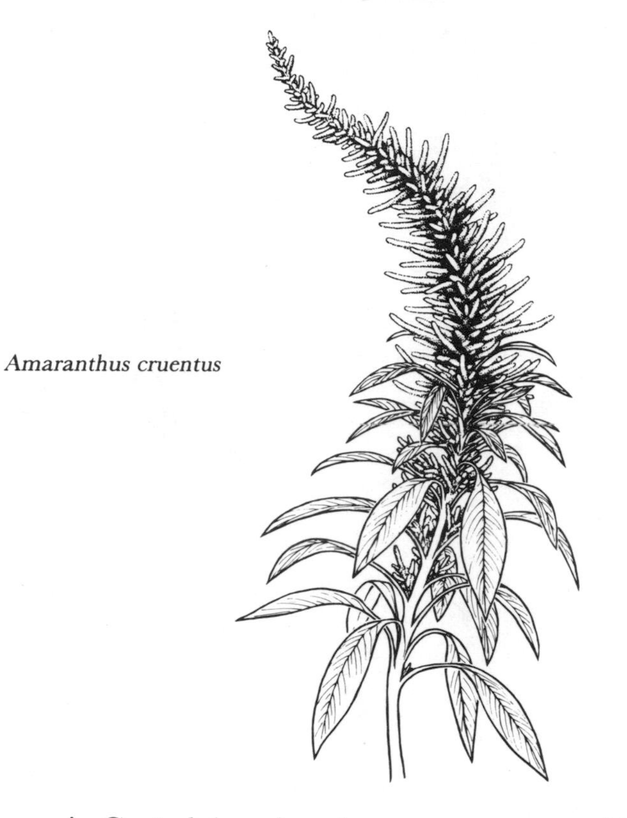

Amaranthus cruentus

crop in Central America, the greens as a vegetable in Africa, and the plants as ornamentals in Europe.

Seed Source:

Not available commercially.

Amaranthus caudatus

Thin spikes are loosely arranged to form the flower head. Individual flowers are extremely crowded on the spikes, sometimes even forming ball-like structures from crowding. When flowering is advanced, the flower heads hang down. Flower color is often red and occasionally green. Leaves are elliptical,

Amaranthus caudatus

tapering toward both ends. Height is generally 5 to 7 feet. Although sometimes light- or dark-colored, *A. caudatus* seeds are characteristically white with a distinctive pink rim.

Use:

A. caudatus is most popularly known in the United States and Europe as the ornamental, love-lies-bleeding. It has been used, however, as a grain and vegetable in India and other Asian and African countries.

Seed Source:

As an ornamental, *A. caudatus* is available from Thompson and Morgan, PO Box 100, Farmingdale, NJ 07727; G. W.

Park Seed Co., PO Box 31, Greenwood, SC 29647; Le Jardin du Gourmet, West Danville, VT 05873; and J. L. Hudson Seedsman, PO Box 1058, Redwood City, CA 94064.

An edible crop of *A. caudatus* is not available commercially.

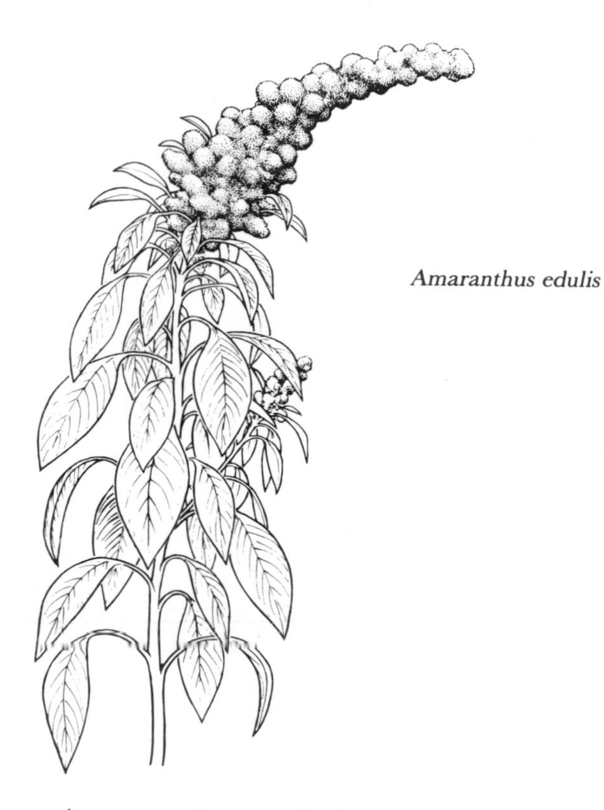

Amaranthus edulis

Amaranthus edulis

The distinctive aspect of *A. edulis* is that the flowers are in ball-like clusters arranged along the spike. The flowers are generally rust colored. Otherwise, this variety is similar in size, habit, and leaf shape to *A. caudatus.*

Use:

Used as a grain and occasionally as a potherb.

Seed Source:

Not available commercially.

Amaranthus retroflexus

Amaranthus retroflexus

A well-known North American weed, pigweed is characterized by gray green hair inflorescences in a sparse flower head. The plants generally reach 2 to 4 feet; the seeds are black. The leaves are oval pointed.

Use:

Primarily as a potherb.

Seed Source:

The weed is primarily found along roadsides, fields, and waste ground in central and eastern United States. Seed is not available commercially.

Amaranthus tricolor

Amaranthus tricolor

This is a vegetable type of amaranth which has a growth habit similar to spinach. The leaves are generally round and wrinkled and are light green, red, or green with red centers. The flowering pattern is quite different from the grain types in that the flowers form along the leaf axils with a small inflorescence on top.

Use:

Both the green and the red forms are edible. As a rule, the green-leaved forms are preferred. This vegetable is grown and eaten much like spinach.

Seed Source:

A. tricolor is sometimes referred to as hinn choy. It's available from Thompson and Morgan, PO Box 100, Farmingdale, NJ 07727; Grace's Gardens, 22 Autumn Lane, Hackettstown, NJ 07840; T. Sakata, CPO Box Yokohama 11, Yokohama Japan 220-91; and Tsang and Ma International, PO Box 294, Belmont, CA 94002.

Amaranthus gangeticus

Amaranthus gangeticus

Commercially known in the United States as tampala, *A. gangeticus* is a spinachlike vegetable that will flourish throughout warm summer months. In fact, it is sometimes called "summer spinach." The dark-green leaves are elongated and smooth. Like *A. tricolor,* the plant can be cut back (ratooned) several times in the growing season and still produce another crop.

Other types of *A. gangeticus* vary widely in leaf shape, leaf color, and height. They are grown in Asia and Africa as vegetables.

Use:

As a green leafy vegetable.

Seed Source:

W. Atlee Burpee Co., Warminster, PA 18974; G. W. Park Seed Co., PO Box 31, Greenwood, SC 29647; Redwood Seed Company, PO Box 361, Redwood City, CA 94064; J. L. Hudson Seedsman, PO Box 1058, Redwood City, CA 94064; and Tsang and Ma International, PO Box 294, Belmont, CA 94002.

APPENDIX D

You can get different forms of *A. gangeticus* and other amaranth vegetables through the Taiwan Seed Service, Shin-Shieh, Taichung Prefecture, Taiwan Province, Republic of China.

Appendix E
Comparative Yields of Common Grains and Amaranth

The table below shows that, even in its earliest stages as a modern food crop, amaranth compares favorably in yield with our most advanced cereal crops. One can assume that yield estimates will increase when agronomists find the most efficient cultural techniques to apply to amaranth and improve upon the best-yielding varieties.

1978 YIELDS OF COMMON GRAINS*

	U.S. Average Yield kg./ha.†
Amaranth‡	900-1,900
Barley	2,600
Corn	6,350
Oats	1,870
Rice	5,040
Rye	1,680
Soybeans	807
Wheat	2,130

* Foreign Agriculture Circular: Grains. January 2, 1979
† 2.2 pounds = 1 kilogram (kg.)   2.4 acres = 1 hectare (ha.)
‡ Preliminary yields from research plots at the Organic Gardening and Farming Research Center

Bibliography

Abdi, H., and Sahib, M. K. 1976. Distribution of lysine in different legumes and some species of *Amaranthus* seeds. *Journal of Food Science Technology* (India) 13:237-39.

Agogino, G. A. 1957. Pigweed seeds dated oldest U.S. food grain. *Science Newsletter* 72:325.

Agogino, G. A., and Feinhandler, S. 1957. Amaranth seeds from a San Jose site in New Mexico. *Texas Journal of Science* 9:154-56.

Ahmad, T. 1938. The amaranth borer, *Lixus truncatulus* (F.) and its parasites. *Indian Journal of Agriculture Science* 9(4):609-27.

Allard, H. A. 1968. Promising selections of green amaranths. *Madras Agricultural Journal* 55:253.

Anderson, E. 1967. *Plants, man and life.* Los Angeles/Berkeley: University of California Press.

Argier, B. 1969. *More free-for-the-eating wild foods.* Harrisburg, Pa.: Stackpole Books.

Beam, A., and Simons, M. 1978. Digging up a goddess. *Newsweek,* 6 November 1978:96.

Bhagat, N. R. 1978. Preliminary notes on the survey and collection of *Amaranthus* in Vidarbha region of Maharashtra-Kharif season.

BIBLIOGRAPHY

Bortz, B. 1975*a*. The chenopods are coming. *Organic Gardening and Farming* 25:30–32.

———. 1975*b*. New crops for America. *Organic Gardening and Farming* 22:106–7.

Brand, D. D. 1939. The origin and early distribution of new world cultivated plants. *Agricultural History* 13:109–17.

Bray, W. J. 1975. The economics of leaf protein production. Presented at the 169th ACS National Meeting, 10 April 1975, Philadelphia, Pennsylvania.

Brenan, J. P. M. 1961. *Amaranthus* in Britain. *Watsonia* 4:261–80.

Callen, E. O. 1965. Food habits of some pre-Columbian Mexican Indians. *Economic Botany* 19:335–43.

Carlsson, R. 1977. *Amaranthus* species and related species for leaf protein concentrate production. In *Proceedings of the First Amaranth Seminar,* 29 July 1977, Maxatawny, Pennsylvania.

Clarke, C. B. 1977. *Edible and useful plants of California.* Los Angeles/Berkeley: University of California Press.

Claus, P. J., and Weyl, A. 1977. Amaranth cultivation in India. Paper written for Rodale Press.

Cook, O. F. 1904. Food plants of ancient America. *Annual Report of the Smithsonian Institute* 1903:481–97.

Deutsch, J. 1977. Genetic variation of yield and nutritional value in several *Amaranthus* species used as a leafy vegetable. Doctoral thesis, Cornell University, Ithaca, New York.

Downton, W. J. S. 1973. *Amaranthus edulis:* a high lysine grain amaranth. *World Crops* 25(1):20.

Dressler, R. L. 1953. The pre-Columbian cultivated plants of Mexico. *Botanical Museum* leaflet, Harvard University 16(6):115–72.

BIBLIOGRAPHY

Early, D. 1977*a*. Amaranth secrets of the Aztecs. *Organic Gardening and Farming* 24:69-73.

———. 1977*b*. Cultivation and uses of amaranth in contemporary Mexico. In *Proceedings of the First Amaranth Seminar*, 29 July 1977, Maxatawny, Pennsylvania.

———. 1977*c*. In search of amaranth. Paper written for Rodale Press.

———. 1978. *Hautli:* The revival of Aztec amaranth: an appropriate technology food. Presented at the 77th Annual Meeting of the American Anthropological Association, Los Angeles, California.

Elias, Joel. 1977*a*. The nutritional value of amaranth. Paper written for Rodale Press.

———. 1977*b*. Food value of amaranth greens and grains. In *Proceedings of the First Amaranth Seminar*, 29 July 1977, Maxatawny, Pennsylvania.

El-Sharkaway, M. A.; Loomis, R. S.; and Williams, W. A. 1968. Photosynthetic and respiratory exchanges of carbon dioxide by leaves of the grain amaranths. *Journal of Applied Ecology* 5:243-51.

Erwin, A. T. 1934-35. Alegriá — a popping seed used in Mexico as a substitute for popcorn. *Iowa State College Journal of Science* 9:661-65.

Feine, L. 1976. The cultivation and domestication of the grain amaranths and their possible use as a future world crop. Paper prepared for Ethnoecology 402, University of Colorado, Boulder, Colorado.

Feine, L. et al. 1979. Amaranth: gentle giant of the past and future. *New Agricultural Crops.* Edited by G. A. Ritchie. A.A.A.S. Selected Symposium #38.

Ford, B. 1978. *Future food: alternate protein for the year 2000.* New York: William Morrow and Company.

Grubben, G. J. H. 1971. L'Amarante et sa culture au Dahomey. *Agronomie Tropicale* 29:97-102.

———. 1976. *The cultivation of amaranth as a tropical leaf vegetable with special reference to South-Dahomey.* Communication 67, Dept. of Agricultural Research, Koninkliijk Instituut voor de Tropen (Royal Tropical Institute), Amsterdam.

Hagen, T. 1960. Afoot in roadless Nepal. *National Geographic* 117(3):361-405.

Halpin, A. 1978. *Unusual vegetables.* Emmaus, Pennsylvania: Rodale Press.

Haughton, C. S. 1978. *Green immigrants: the plants that transformed America.* New York and London: Harcourt Brace Jovanovich.

Hauptli, H. 1977*a*. Horticultural considerations of grain amaranth. Paper written for Rodale Press.

———. 1977*b*. Agronomic potential and breeding strategy for grain amaranths. In *Proceedings of the First Amaranth Seminar,* 29 July 1977, Maxatawny, Pennsylvania.

———. 1978. Biosystematics and agronomic evaluation of some weedy and cultivated amaranth populations. Thesis, University of California.

Hauptli, H. and Jain, S. 1977. Amaranth and meadowfoam: two new crops? *California Agriculture* 31(9):6-7.

———. 1978. Biosystematics and agronomic potential of some weedy and cultivated amaranths. *Theory of Applied Genetics* 52:177-85.

Heiser, C. B., Jr. 1964. Sangorache, an amaranth used ceremonially in Ecuador. *Amaranth Anthropologist* 66:136-39.

Herklots. G. A. C. 1972. *Vegetables in Southeast Asia.* London: George Allen and Unwin.

Howell, J. T. 1933. The Amaranthaceae of the Galapagos Islands. *Proceedings of the California Academy of Sciences* 21(9):87-116.

Jain, S. 1978. Correspondence to Nancy Nickum Bailey from sabbatical leave travels in India, 24 December.

Kamalanathan, S. et al. 1970. Studies on the optimum time of sowing and stage of harvest of Co. 1, *Amaranthus (A. flavus* L.). *South Indian Horticulture* 18(314):77-80.

Khoshoo, T. N., and Pal, M. 1972. Cytogenetic patterns in *Amaranthus. Chromosomes Today* 3:259-67.

Knott, J. E., and Deanon, J. R., Jr. 1967. *Vegetable production in Southeast Asia.* Los Banos, Laguna, Philippines: University of the Philippines.

Krochmal, A.; Paur, S.; and Duisberg, P. 1954. Useful native plants in the American southwestern deserts. *Economic Botany* 8(1):3-20.

Kurien, P. P. 1967. Distribution of protein, calcium and phosphorus between the husk and endosperm of rajgira seeds (*Amaranthus paniculatus*). *Journal of Nutrition and Dietetics* (India) 4:153-55.

MacNeish, R. S. 1964. Ancient Mesoamerican civilization. *Science* 143(3606):531-37.

———. 1967. A summary of subsistence. *Environment and subsistence.* The Prehistory of the Tehuacán Valley, edited by D. S. Byers, vol. 1. Austin: University of Texas Press.

———. 1971. Speculation about how and why food production and village life developed in the Tehuacán Valley, Mexico. *Archaeology* 24(4):307-15.

Martin, F. W., and Ruberte, R. M. 1975. *Edible leaves of the tropics.* Mayaguez: Antillian College Press.

———. 1977. Selected amaranth cultivars for green leaves. In *Proceedings of the First Amaranth Seminar,* 29 July 1977, Maxatawny, Pennsylvania.

BIBLIOGRAPHY

Marx, J. L. 1977. Amaranth: A comeback for the food of the Aztecs? *Science* 198(4312):40.

Mathai, P. J. 1978. *Amaranthus:* a neglected vegetable. *Indian Farming* 28(1):29,32.

Menninger, R. 1977. New crops. *Co-Evolution Quarterly* 15:76-79.

Messer, E. 1972. Patterns of "wild" plant consumption in Oaxaca, Mexico. *Ecology of Food and Nutrition* 1:325-32.

Mohideen, M. K., and Rajagopal, A. 1975. Effect of transplanting on growth, flowering and seed yield in *Amaranthus. South Indian Horticulture* 23(3/4):87-90.

Munger, H. M., and Deutsch, J. A. 1977. Amaranth as a leafy vegetable. Paper written for Rodale Press.

Nabhan, G. P. 1978. Warihio Indians and amaranth agricul ture in the northern Sierra Madre. University of Arizona, Plant Sciences Department, Tucson, Arizona.

National Academy of Sciences 1975. *Underexploited tropical plants with promising economic value.* Washington, D.C.: National Academy of Sciences.

Ochse, J. J. 1931. *Vegetables of the Dutch East Indies.* Buitenzorg.

Pal, M. 1971. Evolution and improvement of cultivated amaranths. 2. A polyhaploid plant of *Amaranthus dubius. Indian Journal of Genetics* 31(3):397-402.

———. 1972*a*. Evolution and improvement of cultivated amaranth. 1. Breeding system and inflorescence structure. *Proceedings of the Indian National Science Academy* 38:B(1&2):28-37.

———. 1972*b*. Evolution and improvement of cultivated amaranths. 3. *Amaranthus spinosus-dubius* complex. *Genetica* 43(1):106-18.

Pal, M., and Khoshoo, T. N. 1972*a*. Evolution and improvement of cultivated amaranths. 4. Variation in pollen mitosis in the F1 *Amaranthus spinosus* x *A. dubius. Genetica* 43(1):119-29.

———. 1972*b*. Evolution and improvement of cultivated amaranths. 5. Inviability, weakness and sterility in hybrids. *Journal of Heredity* 63:78-82.

———. 1973*a*. Evolution and improvement of cultivated amaranths. 6. Cytogenetic relationships in grain types. *Theory of Applied Genetics* 43:242-51.

———. 1973*b*. Evolution and improvement of cultivated amaranths. 7. Cytogenetic relationships in vegetable amaranths. *Theory of Applied Genetics* 43:343-50.

———. 1974. Grain amaranths. *Evolutionary studies in world crops: diversity and change in the Indian subcontinent.* Edited by J. B. Hutchinson. New York: Cambridge University Press.

———. 1977. Evolution and improvement of cultivated amaranths. 8. Induced autotetraploidy in grain types. *Z. Pflanzenzüchtg* 78:135-48.

Pirie, N. W. 1966. Leaf protein as human food. *Science* 152(3730):1701-5.

Powers, W. 1976. Correspondence to Nancy Nickum Bailey, 4 December.

Raker, D. S. 1978. The modern amaranth: a review of current literature on a potentially important food crop. Paper prepared for Biology 104, Harvard University, Cambridge, Massachusetts.

Robson, J. R. K. 1973. Correspondence to Robert Rodale, 28 June.

——— 1977*a*. Amaranth and contemporary food habits. Paper written for Rodale Press.

———. 1977*b*. Value of indigenous food (abstract). In *Pro-*

ceedings of the First Amaranth Seminar, 29 July 1977, Maxatawny, Pennsylvania.

Robson, J. R. K., and Elias, J. N. 1978. *The nutritional value of indigenous wild plants: an annotated bibliography.* Troy, New York: Whitston Publishing Company.

Rodale, R. 1974. The half-wild way to a better life. *Organic Gardening and Farming* 21:38–42.

———. 1976. Goodbye red no. 2, hello amaranth. *Prevention* 28:23-28.

———. 1978. Correspondence to Dr. David E. Walsh, 15 December.

———. 1979. The secret health of plants. *Organic Gardening* 26(2):36–42.

Rosenthal, J. E. 1978. In search of hearty crops. *The Christian Science Monitor,* 6 July, p. 13.

Ruttle, J. 1976. Amaranth the gentle giant. *Organic Gardening and Farming* 23:106–10.

Safford, W. E. 1915. A forgotten cereal of ancient America. *Proceedings of the 19th International Congress of Americanists,* Washington, D.C.

Sant, V. Hunger and malnutrition: a Hiroshima every 3 days. . . . *A Shift in the Wind* 2:45.

Sauer, J. D. 1950*a*. Amaranths as dye plants among Pueblo peoples. *Southwestern Journal of Anthropology* 6:412–15.

———. 1950*b*. The grain amaranths: a survey of their history classification. *Annals of the Missouri Botanical Garden* 37:561–619.

———. 1953. Herbarium specimens as records of genetic research. *American Naturalist* #87(834):155, 156.

———. 1955. Revision of the dioecious amaranths. *Madroño* 13:5–46.

———. 1957. Recent migration and evolution of the dioecious amaranths. *Evolution* 11:11-31.

———. 1967. The grain amaranths and their relatives: a revised taxonomic and geographic survey. *Annals of the Missouri Botanical Garden* 54(2):103-37.

———. 1969. Identity of archaeologic grain amaranths from the valley of Tehuacán, Puebla, Mexico. *American Antiquity* 34(1):80-81.

———. 1976. Grain amaranths, *Amaranthus* spp., (Amaranthaceae) *Evolution of Crop Plants.* Edited by N. W. Simmond. 2:4-6.

———. 1977*a*. The history of grain amaranths and their use and cultivation around the world. In *Proceedings of the First Amaranth Seminar,* 29 July 1977, Maxatawny, Pennsylvania.

———. 1977*b*. The history of amaranth use and cultivation around the world. Paper written for Rodale Press.

Schmidt, D. R. 1971. Comparative yields and composition of eight tropical leafy vegetables grown at two soil fertility levels. *Agronomy Journal* 63(4):546-50.

———. 1977. Grain amaranth: a look at some potentials. In *Proceedings of the First Amaranth Seminar,* 29 July 1977, Maxatawny, Pennsylvania.

Scullin, M. 1968. The grain amaranths and chenopods as components of South American agriculture. Paper prepared for Anthropology 429.

Singh, H. 1961. Grain amaranths, buckwheat and chenopods. *Indian Council of Agricultural Research.* Cereal Crop Series No. 1, New Delhi, India.

Smith, C. E., Jr. 1965. The archaeological record of cultivated crops of new world origins. *Economic Botany* 19(4):322-34.

———. 1967. Plant remains. *Environment and subsistence.*

The Prehistory of the Tehuacán Valley, edited by D. S. Byers, vol. 1. Austin: University of Texas Press.

Srinivasan, A. R. 1978. Rich in nutrition, but little exploited. *The Hindu.* Madras, India.

Stafford, W. L.; Mugerwa, J. S.; and Bwabye, R. 1976. Effects of methods of cooking, application of nitrogen fertilizer and maturity on certain nutrients in the leaves of *Amaranthus hybridus* subspecies *hybridus* (Green Head). *Plant Foods for Man* 2:7-13.

Standley, P. C. 1912. Some useful native plants in New Mexico. *Smithsonian Institution, Annual Report of the Board of Regents for 1911.* pp. 447-62.

———. 1917. *Amaranthus. North American Flora* 21:99-119.

———. 1937. Amaranthaceae — flora of Peru. *Chicago Field Museum of Natural History — Botany,* 13:478-518.

Tompkins, P. 1976. *Mysteries of the Mexican Pyramids.* New York: Harper & Row.

Tucker, J. M., and Sauer, J. D. 1958. Aberrant *Amaranthus* populations of the Sacramento-San Joaquin Delta, California. *Madroño* 14:252-61.

Van Etten, C. H. et al. 1963. Amino acid composition of seeds from 200 angiospermous plant species. *Journal of Agricultural and Food Chemistry* 11(5):399-410.

Vietmeyer, N. D. 1978. The plight of the humble crops. *Ceres* 11(2):23-27.

Walker, H. G., Jr. et al. 1970. Preparation and evaluation of popped grains for feed use. *Cereal Chemistry* 47:513-21.

Walton, P. D. 1968. The use of the genus *Amaranthus* in genetic studies. *Journal of Heredity* 59:76.

Whitaker, T. W., and Cutler, H. C. 1966. Food plants in a Mexican market. *Economic Botany* 20(1):6-16.

White, G. A., and Wolff, I. A. 1968. From wild plants to new crops in USA. *World Crops* 20(3):70-76.

White, G. A. et al. 1971. Agronomic evaluation of prospective new crop species. *Economic Botany* 25:22-43.

Wilkes, G. 1977. Native crops and wild food plants. *The Ecologist* 7(8):312-17.

Zabka, G. C. 1961. Photoperiodism in *Amaranthus caudatus.* 1. A re-examination of the photoperiodic response. *American Journal of Botany* 48:21-28.

Index

Asterisks (*) indicate recipes

A

INDEX

D

R

S

T